THE COMMONWEALTH AND INTERNATIONAL LIBRARY

Joint Chairmen of the Honorary Editorial Advisory Board
SIR ROBERT ROBINSON, O.M., F.R.S., LONDON
DEAN ATHELSTAN SPILHAUS, MINNESOTA

Publisher: ROBERT MAXWELL, M.C., M.P.

METALLURGY DIVISION
General Editors: W. S. OWEN; D. W. HOPKINS; H. M. FINNISTON

The Kinetics of Phase
Transformations in Metals

The Kinetics of Phase Transformations in Metals

BY

J. BURKE

Professor of Physical Metallurgy, University College, Swansea

PERGAMON PRESS

OXFORD · LONDON · EDINBURGH · NEW YORK
PARIS · FRANKFURT

Pergamon Press Ltd., Headington Hill Hall, Oxford
4 & 5 Fitzroy Square, London W.1
Pergamon Press (Scotland) Ltd., 2 & 3 Teviot Place, Edinburgh 1
Pergamon Press Inc., 44–01 21st Street, Long Island City, New York 11101
Pergamon Press S.A.R.L., 24 rue des Écoles, Paris 5e
Pergamon Press GmbH, Kaiserstrasse 75, Frankfurt-am-Main

Copyright © 1965 Pergamon Press Ltd.
First Edition 1965
Library of Congress Catalog Card No. 65–26345

Printed in Great Britain by Bell & Bain Ltd., Glasgow

This book is sold subject to the condition
that it shall not, by way of trade, be lent,
resold, hired out, or otherwise disposed
of without the publisher's consent,
in any form of binding or cover
other than that in which
it is published.

(2269/65)

Contents

	Preface	vii
1.	The Basis of Kinetic Theory	1
2.	Empirical Kinetics	36
3.	Diffusion in Metals	61
4.	Phase Changes in Metals and Alloys	92
5.	Nucleation	98
6.	Theory of Diffusional Growth Processes	152
7.	The Kinetics of Diffusional Transformations	184
8.	Martensitic Transformations	196
	Appendix A	214
	Appendix B. Units and Symbols	220
	Index	223

Preface

THE kinetics of phase transformations is studied in many university courses in metallurgy and materials science. In view of the considerable experimental and theoretical activity in this field over the last twenty years it is surprising that there is an acute shortage of text books suitable for students. The relevant material is spread over a number of texts in physics and physical chemistry together with papers and reviews in a variety of journals. Many of these assume a greater knowledge of statistical thermodynamics and mathematics than most undergraduate students of metallurgy ordinarily possess and others are too specialized to be suitable for students meeting the subject for the first time. In this book an attempt has been made to present a concise introduction to the basic principles of kinetics and to show how these have been applied to some problems of current interest so that students may then be able to study, with profit, the more advanced texts and papers. The mathematics are kept to an elementary level throughout and the only prior knowledge assumed is of thermodynamics and metallography.

The book was developed from a course of lectures to the final year of an Honours Degree Course in Metallurgy given in the Metallurgy Department, Liverpool University. In the first chapter some basic principles are presented with particular emphasis on the concept of an activation energy. In the second chapter some of the empirical ideas that have proved generally useful are discussed and the methods of deriving empirical parameters from experimental data explained. The kinetics of some specific processes of metallurgical importance are treated in the remaining chapters. These chapters are not intended to be comprehensive or critical accounts of the present state of knowledge in the various fields; rather are they intended to be simple accounts suitable for students interested in gaining some familiarity with a range of topics. Thus the treatments in most cases are oversimplified and many difficulties are glossed over. Recent authoritative reviews are listed at the end of each chapter.

It is a pleasure to acknowledge the valuable help and advice received from Professor W. S. Owen during the preparation of the book. I am also grateful to Dr. R. D. Townsend who read the whole draft manuscript and made many suggestions for improvement.

CHAPTER 1

The Basis of Kinetic Theory

1.1. Stable and Metastable Equilibrium

A system at constant temperature T and pressure P is in equilibrium when its Gibbs free energy G is a minimum. Thus an equilibrium configuration of a system is characterized by

$$dG_{T,P} = 0 \qquad (1.1)$$

in which dG represents the change in G associated with an infinitesimal change in the system, and the subfixes denote the variables held constant. dG is given by

$$dG_{T,P} = dU + PdV - TdS \qquad (1.2)$$

in which U is the internal energy, V the volume and S the entropy. The changes in volume accompanying changes of state in condensed systems are normally very small and may be neglected permitting the use of the approximations

$$dG \simeq dF \qquad (1.3)$$
$$dU \simeq dH \qquad (1.4)$$

F is the Helmholtz free energy and H the enthalpy, which are defined by the relations

$$dF = dU - TdS \qquad (1.5)$$
$$dH = dU + PdV \qquad (1.6)$$

The equilibrium condition, eqn. (1.1) then becomes

$$dF_{T,P} = 0 \qquad (1.7)$$

Changes in G (and H) are more easily measured experimentally than are changes in F, because most experiments are performed at constant pressure rather than at constant volume. Theoretically it is more convenient to deal with constant volume conditions and thus with F and U. The use of the approximations, eqns. (3) and (4), enables theory and experiment to be compared directly. This procedure will be used throughout.

Equilibrium may be stable or metastable. A system is in *stable equilibrium* when its free energy is the lowest value possible consistent with the imposed external conditions. Configurations of the basic particles of the system for which G is a minimum in the mathematical sense, satisfying eqns. (1) and (7), but numerically greater than the value of G associated with the stable configuration, are *metastable*.

Only stable or metastable configurations are ever realized in practice. If, by some means, a system is produced in some other state, any slight disturbance (always present in the form of thermal fluctuations) reduces G. Hence the change is favourable and will continue until the system reaches an adjacent equilibrium state. In contrast, in stable or metastable states, small fluctuations increase G, providing a restoring force opposing the change and so slight changes so produced have only transient existence.

1.2. Definition and Classification of Transformations

Any rearrangement of the atoms, ions or molecules of a system from one metastable configuration to another of lower free energy is referred to as a transformation, reaction or transition. The product of a transformation need not be the stable configuration; it may be a second metastable state of lower free energy than the first. It is then possible for this product to undergo a further transformation to a more stable configuration, and this sequence may be repeated until the stable form is achieved. An example is the precipitation of supersaturated copper from quenched aluminium —4% Cu alloys, which at room temperature and slightly above, occurs in four consecutive stages: (*a*) the formation of Guinier-Preston (1) zones—copper rich clusters; (*b*) the conversion of G.P. (1) zones into G.P. (2) zones—an ordered form of G.P. (1); (*c*) the production of θ', a metastable tetragonal phase, at the expense of the zones, and finally (d) the formation of the equilibrium phase, tetragonal θ, $CuAl_2$. Although each stage is essentially a separate reaction, in practice it is usual for the various stages of multi-stage processes to overlap, and the interdependence of each stage makes a quantitative understanding far more difficult than in the case of a single stage change.

A transformation may involve the complete rearrangement of every particle in the system as when b.c.c. α-Fe changes to f.c.c. γ-Fe during heating through the A_3 temperature; or only a fraction of the particles may be affected as during the precipitation of excess solute atoms from supersaturated solutions.

The macroscopic result of a transition is achieved by a large number of repetitions of one or more basic atomic processes. Comparison of the experimentally observed kinetics with those derived on the basis of assumed models is a powerful means of elucidating the particular atomic processes involved in a transformation. The elementary processes vary from the movement of single atoms, as in diffusion, to quite complicated mechanisms involving the co-operative action of many atoms. The easiest cases to treat theoretically are those involving only one step, but it is difficult to realize these " ideal " conditions in practice. For example, it is reasonably certain that in many metals diffusion through the lattice occurs by individual atoms jumping into neighbouring vacant sites. However, in an actual diffusion experiment diffusion along the surface, along grain boundaries and along dislocations also contribute to the observed flow. Most real reactions involve a number of consecutive or simultaneous basic atomic events and the difficulties and uncertainties in the treatment of them is considerable.

Homogeneous reactions are those which occur within a single phase. Superficially all gas reactions are of this type. In fact most are found to be heterogeneous in that most of the reaction is accomplished at the container walls. The most important homogeneous reaction in solid state metallurgy is true lattice diffusion. *Heterogeneous* reactions involve two or more phases and reaction occurs at the phase boundary. At least two consecutive atomic steps are required: (a) transport of atoms to the phase boundary and (b) reaction at the boundary. Phase changes, precipitation and dissolution of solute, re-crystallization, solid–gas and solid–liquid reactions are all heterogeneous.

1.3. Definition of Reaction Rate

All the particles in a system do not undergo transformation at one and the same time. If they did then at any instant during the reaction all the particles would be in some intermediate configuration and neither the initial nor the final configurations would be detectable. Observation shows that reactants and products co-exist throughout the transformation. Indeed, the most careful experiments fail to reveal the presence of any other component.† Evidently, at any instant, an immeasurably small fraction of the total particles available for reaction are actually in the process of change and the vast majority are either completely transformed or completely untransformed. The reason for this is discussed in section 1.6. Hence the extent to which a reaction has proceeded at a given time t may be described by the fractional transformation $y(t)$ defined as the ratio of the number of atoms per unit volume in the final configuration at time t, to the number per unit volume available for transformation at $t = 0$. The rate of reaction at time t is then simply dy/dt. In general, dy/dt is a function of time.

1.4. The Driving Force for a Transformation

It follows from the second law of thermodynamics that the driving force for a transformation is the difference between the free energy of the initial state G_I and that of the final state G_F. $\Delta G = (G_F - G_I)$ must be negative for a change to have a positive driving force.

All metastable states have a finite driving force tending to

---

† For those changes such as the precipitation of copper from aluminium referred to in previous paragraphs which proceed via a sequence of sub-reactions the products of the various stages are not intermediate configurations in this sense, since each stage is a separate reaction.

produce transformation to a more stable state. It is a matter of practical experience that despite this, many systems are capable of existing indefinitely in metastable forms. For example, quenching a carbon steel from a temperature above that of the pearlite-austenite transformation produces a mixture of austenite and martensite, which is metastable relative to a mixture of ferrite and cementite; and yet the quenched structure will remain unchanged indefinitely at room temperature. Whilst a finite driving force is a necessary condition for a particular transformation to occur, it is evidently no guarantee that it will proceed at a sensible rate. In fact in a wide range of reactions there is no correlation between the driving force and the rate of the reaction. This point is well illustrated by the influence of catalysts. The balance of a chemical equilibrium is not changed by the presence of a catalyst showing that the free energy change accompanying a reaction (i.e. the driving force) is not altered. In other words, the acceleration of the reaction rate is accomplished without alteration of the driving force. It is to be concluded that the kinetics of a reaction are determined largely by factors other than the driving force. To isolate these factors it is necessary to consider the states through which a system passes during transformation.

1.5. The Usefulness of Equilibrium Thermodynamics

The fact that the terminal states of a transformation are equilibrium configurations permits the application of normal thermodynamics and statistical mechanics to these states. The configurations through which a system passes during transformation cannot, strictly, be described by these methods. Two attempts have been made to overcome this fundamental difficulty: (*a*) to treat a transformation as a multi-body problem of particle dynamics, and (*b*) to develop a system of irreversible thermodynamics which includes equilibrium as a limiting condition. Application of these methods to relatively complex heterogeneous reactions has so far not been attempted. Consequently these theoretical treatments will not be considered. The alternative is a quasi-equilibrium approach.

As a system transforms from the initial state to the final state it passes through a continuous series of intermediate configurations. In the quasi-equilibrium approach it is assumed that one of these intermediate states—*the transition state*—is a quasi-equilibrium state and thus has unique values of the thermodynamic functions. The kinetic problem, strictly outside the scope of normal thermodynamics, is reduced to an analysis of the equilibrium between atoms in the initial and transition configurations. Despite its limitations this concept has been the basis of much of the modern thinking about reaction kinetics. All the theories presented in this book are of this type. An elementary partial justification is given in section 1.11.

1.6. The Transition or Activated State—the Free Energy of Activation

The first step in this quasi-equilibrium approach is to define the transition state. As an atom moves from an initial equilibrium state to a final one it passes through a continuous sequence of intermediate states. Since the free energies of the two extreme configurations are, by definition, minima and two minima must be separated by a maximum, it follows that the free energy of an atom or group of atoms during transformation first increases to a maximum and then decreases to the final value. This is illustrated in Fig. 1.1. G_I is the mean free energy of an atom in the initial configuration, and G_F is that after transformation. $\Delta G = (G_F - G_I)$ is negative and is the driving force for the change. An atom having the maximum free energy G^*_A is unstable, being able to either revert to the initial state or proceed to the final state with a reduction in free energy. The configuration associated with this maximum in the free energy curve is assumed to be the *transition* or *activated* state.

Evidently a necessary condition for an atom to take part in the change is that it has sufficient free energy to enable it to achieve transition status; i.e. its free energy relative to the average value in the initial state must be not less than $G_A = (G^*_A - G_I)$. G_A is known as the *free energy of activation* for the reaction.

THE BASIS OF KINETIC THEORY

The additional free energy necessary for an atom to surmount this thermodynamic barrier to transformation is supplied by thermal fluctuations. The distribution of energy amongst an assembly of particles is not uniform. At all temperatures above 0°K the particles are in motion. The collisions which result from this random motion produce wide variations in the energy of individual particles and fluctuations with time in the energy of any single particle. At any instant the assembly embraces a wide spectrum of energies and some particles have energies greatly in excess of the mean. Those with excess free energy equal to, or greater than, G_A will transform. Those with insufficient free energy must wait until they receive the necessary activation energy from thermal fluctuations. The process is termed *thermal activation*.

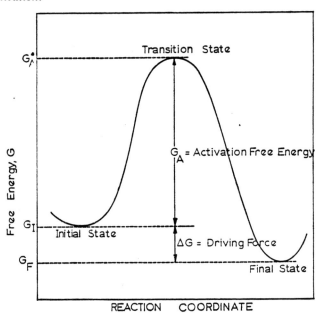

FIG. 1.1. The change in free energy of an atom as it takes part in a transition. The "reaction coordinate" is any variable defining the progress along the reaction path.

It is to be expected from these ideas that reaction velocities will depend very much upon the magnitude of the activation free energy and upon the form of the energy distribution resulting from the random thermal motion. However, before discussing these points further, it is appropriate to note that the concept of an activation energy barrier enables several well-known features of reactions to be explained qualitatively. It is now obvious that at any instant only a small fraction of the available particles can be in the process of transformation for the simple reason that only a fraction have a free energy in excess of the mean. The persistence of metastable states is due to the activation free energy being very large in comparison with the mean free energy, so that the probability of an atom gaining sufficient free energy from thermal fluctuations is nearly zero. The previously mentioned independence of the velocity of reactions and the driving force is also understandable. It can be seen from Fig. 1.1 that ΔG and G_A are not related. For example, it is possible to draw many other curves starting from G_I and ending at G_F all with different activation free energies. Thus it is possible to alter the kinetics of a reaction, through G_A, without changing the thermodynamics. A catalyst acts by providing a reaction path of lower activation free energy.

In introducing the concept of activation free energy it was implied that for a given reaction there is a unique sequence of intermediate configurations. In fact it is always possible to find more than one set of configurational changes capable of producing a given transformation. For example, the basic step in self-diffusion is the movement of atoms from one site to another. Three possible ways of realizing this are: (*a*) exchange with a vacant site, (*b*) direct interchange with a neighbour and (*c*) movement through interstitial space. Each of these "reaction paths" gives a free energy curve similar in form to Fig. 1.1, but, because different transition configurations are involved, the activation free energies are different. For energetic reasons the path actually realized is that associated with the lowest activation free energy which in this example is (*a*). In general the free energy curves associated

THE BASIS OF KINETIC THEORY

with all the possible reaction paths, when assembled together, form a free energy surface. This surface contains a saddle point; the configuration corresponding with the saddle point is the transition state. The free energy of the saddle point configuration is smaller than that of all adjacent configurations save two; one of which leads to the initial state and the other to the final state. Using a geographical analogy, the actual reaction path is the pass across a range of energy hills. The free energy change along this path is the one plotted in Fig. 1.1.

1.7. Activation Energy and Entropy

Although the concept of an activated state arises naturally from the definition of equilibrium in terms of G, it is often more convenient to discuss the thermodynamics of this state in terms of an (internal) energy of activation U_A and an entropy of activation S_A. U_A and S_A are related to G_A by the standard equation†

$$G_A = U_A - TS_A \tag{1.8}$$

The *activation energy* U_A is defined as the difference between the internal energy of an atom in the transition state and one in the initial state. For the present purposes the internal energy of a system of atoms may be considered as divided into two components: (*a*) the potential energy of interaction of the atoms associated with the binding forces, and (*b*) the kinetic energy of the thermally induced motion of the particles. To a good approximation these two are independent and changes in the internal energy may be obtained by evaluating the changes in the two components separately.

---

† For strictly consistent symbolism the activation free energy, energy and entropy should be written ΔG_A, ΔU_A and ΔS_A to emphasize that they are differences between the activated and initial state. However, the symbol Δ will be restricted to changes involved in the complete reaction and free symbols to the activation stage.

10   THE KINETICS OF PHASE TRANSFORMATIONS IN METALS

Considering first the interaction energy, mechanical equilibrium requires that the atoms in the initial and final configurations occupy positions of minimum potential energy. The potential

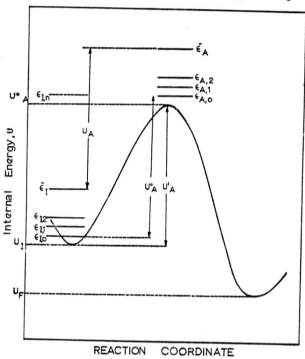

FIG. 1.2. An illustration of the relationship between the activation energies U_A^1, U_A^0 and U_A. U_1, U_A^* and U_F are the potential energies of an atom in the initial, transition and final states respectively. The thermal energy levels for the initial and transition states are shown by the horizontal lines and numbered $\epsilon_{I,0}, \epsilon_{I,1}, \ldots, \epsilon_{A,0}, \epsilon_{A,1}, \ldots, \bar{\epsilon}_I$ and $\bar{\epsilon}_A$ are the mean thermal energies in the two states at some arbitrary temperature.
The curve is drawn for an exothermic process.

energy of an atom during transformation must, therefore, trace out a curve similar to that of the free energy curve in Fig. 1.1. There is one important difference. U_F may be smaller or greater than U_I whereas G_F must always be less than G_I. If ΔU is positive the reaction is endothermic; if ΔU is negative it is exothermic. The potential energy of the transition state relative to that of the initial state is U_A^1. This is the activation energy for the change if thermal energy is neglected (see Fig. 1.2). Obviously it must always be positive. In principle, the value of U_A^1 for a particular process can be calculated from electron theory.

Turning now to the thermal energy, according to the quantum theory only discrete energy levels are permitted. Suppose that the levels for the thermal energy of atoms in the initial state are $\varepsilon_{I,0}, \varepsilon_{I,1}, \ldots, \varepsilon_{I,i}$ and for the activated state $\varepsilon_{A,0}, \varepsilon_{A,1}, \ldots, \varepsilon_{A,i}$. $\varepsilon_{I,0}$ is the *ground* or *zero* point energy level and is the thermal energy which every particle of the system has at 0°K. Quantum theory permits ε_0 to have finite values. Since the thermal energy and the interaction energy are additive the ε's must be measured from the potential energy of the state as zero, as shown in Fig. 1.2. The true difference, at 0°K, between the internal energy of an atom in the transition state and in the initial state, denoted by U_A^0 is obtained by adding the difference between $\varepsilon_{A,0}$ and $\varepsilon_{I,0}$ to U_A^1, i.e.

$$U_A^0 = U_A^1 + (\varepsilon_{A,0} - \varepsilon_{I,0}) \tag{1.9}$$

U_A^0 is called the *zero point activation energy* and is the activation energy that the reaction would have if it could take place at 0°K.

As the temperature is raised the thermal energy of the assembly of particles increases and atoms are excited into higher energy levels. At a given temperature every atom in a system is not in the same level. Due to thermal fluctuations the atoms are distributed amongst the permitted levels. Let the mean thermal energy per atom in the initial state at some arbitrary temperature T be $\bar{\varepsilon}_I$. Similarly, the atoms in the transition state are distributed amongst their permitted energy levels. Let the mean thermal energy of the activated state be $\bar{\varepsilon}_A$. Then it is evident from Fig. 1.2 that the activation energy at that temperature, being the difference between the mean internal energy of the transition and initial state is

$$\begin{aligned} U_A &= U_A^1 + (\bar{\varepsilon}_A - \bar{\varepsilon}_I) \\ &= U_A^0 + (\bar{\varepsilon}_A - \varepsilon_{A,0}) - (\bar{\varepsilon}_I - \varepsilon_{I,0}) \end{aligned} \tag{1.10a}$$

($\bar{\varepsilon}_I - \varepsilon_{I,0}$), the energy absorbed in raising the temperature from $0°K$ to T is equal to

$$\int_0^T C_{V(I)} \cdot dT$$

where $C_{V(I)}$ is the specific heat at constant volume for the initial state. Similarly,

$$(\bar{\varepsilon}_A - \varepsilon_{A,0}) = \int_0^T C_{V(A)} \cdot dT$$

and so

$$U_A = U_A^0 + \int_0^T (C_{V(A)} - C_{V(I)}) dT \qquad (1.10b)$$

In most cases there is insufficient evidence about the nature of the activated state to enable $C_{V(A)}$ to be evaluated. For simple processes at least it is reasonable to suppose that $C_{V(A)}$ will differ little from $C_{V(I)}$ and thus the integral in eqn. (1.10b) may be neglected in comparison with U_A^0. Thus, in general,

$$U_A \simeq U_A^0 \qquad (1.10c)$$

The entropy of activation. S_A is the difference between the entropy of an atom in the activated state and in the initial state. From the Boltzmann equation S_A per atom is given by

$$S_A = k \ln W_A/W_I \qquad (1.11)$$

where k is the Boltzmann constant, 1.38×10^{-16} erg/°C and W_A and W_I are the number of complexions associated with the activated state and initial state respectively. The entropy change comprises changes in configurational entropy associated with the change in the spatial distribution of the atoms, changes in thermal entropy due to the different ways in which the thermal energy is distributed over the permitted energy levels, changes in electronic entropy and changes in the distribution of other forms of energy.

It is not possible to generalize about the sign or magnitude of S_A. In principle it is possible to calculate S_A for a given atomic process but in practice this can seldom be done.

Thus the requirement that an atom must have the free energy of activation G_A before it can participate in a reaction is equivalent to the two conditions that (a) an atom must have a thermal (kinetic) energy at least equal to U_A to enable it to overcome the potential barrier, and (b) simultaneously the entropy requirements of the transition configuration are satisfied. Either of these is a necessary but not a sufficient condition for reaction, whereas the possession of the free energy of activation is in itself the necessary and sufficient condition.

In view of these requirements the next stage in developing a kinetic theory is to inquire how the thermal energy is distributed among the particles of a system in equilibrium because this determines the number of atoms having sufficient thermal energy to overcome the potential energy barrier.

1.8. The Distribution of the Thermal Energy of a System amongst its Particles

Each macroscopic state of a system defined by specified values of the macroscopic variable P, T, U, etc., corresponds to a very large number of microscopic states. For example, a gas in equilibrium in a vessel has fixed values of P and T although the distribution of molecules in space, and the distribution of energy amongst the molecules changes continuously. In statistical mechanics each distinguishable microscopic arrangement is known as a *complexion* and the total number of complexions W associated with a given thermodynamic state is the *probability* of that state. For example, in a pure metal at $0°K$, the atoms are at rest on their lattice sites, in the ground energy level, with all sites occupied. Atoms of the same isotope of the same element and having the same energy are indistinguishable and so interchange of two atoms does not give a recognizably different arrangement; thus in this case $W = 1$. If two isotopes are present exchange of the isotopes leads to new complexions. If there are n_1 atoms of one kind and n_2 of the other,

$$W = \frac{(n_1 + n_2)!}{n_1!\, n_2!}$$

A fundamental assumption of statistical mechanics is that every complexion is equally likely to occur. When a system is allowed to come to equilibrium all complexions are explored impartially due to the random atomic motion. All the conceivable macroscopic states have a certain number of complexions. Simple probability considerations lead to the conclusion that in closed systems (i.e. those of constant volume V and internal energy U) the state most likely to be found is that with the greatest W because it can be realized in the greatest number of ways. In other words the equilibrium state is that of maximum probability. In the isotope example two conceivable states are (a) random mixing on the available sites, and (b) complete segregation. The value of W for (b) is unity and for (a) is

$$\frac{(n_1 + n_2)!}{n_1!\, n_2!}$$

The chance of finding (a) relative to that of (b) is also the factorial expression, a very large number for reasonable values of n_1 and n_2. Random mixing is thus the equilibrium arrangement.

W for a given macroscopic state is related to the entropy S by Boltzmann's equation

$$S = k \ln W$$

Hence the equilibrium condition that W is a maximum is equivalent to making S a maximum, which is the condition also required by the second law of thermodynamics.

Entropy may arise in several different ways. Configurational entropy is associated with the different ways of arranging the atoms in space, e.g. the isotope example above. Thermal energy may have translational, vibrational and rotational components with corresponding entropies arising from the different ways of distributing the different types of energy amongst the permitted energy levels. The problem here is to find the most probable way of distributing a total thermal energy U_T of unspecified type, given that the permitted energy levels are $\varepsilon_0, \varepsilon_1, \varepsilon_2, \ldots, \varepsilon_i$. Then according to statistical mechanics this distribution is also the energy distribution existing in a metastable equilibrium state prior to a transformation.

The energy levels are determined by solving Schrödinger's equation for all possible types of energy and then combining the different energy levels in all possible ways to give the combined levels $\varepsilon_0, \varepsilon_1$, etc. It is found that these levels are functions of the volume and so the problem is much simplified if constant volume systems are considered. In such systems the ε's are independent of temperature. Furthermore, having once fixed the volume all thermodynamic properties (e.g. U) are functions of the temperature only.

Consider a system of fixed volume V, temperature T, and thermal energy U_T containing N identical particles, constrained to vibrate about fixed sites in the crystal. One possible arrangement of the atoms consists of n_0 atoms with energy ε_0, n_1 with ε_1, and so on; in general, let there be n_i in the ε_i level. The arrangement must satisfy the requirements

$$\Sigma_i n_i = N \qquad (1.12)$$

$$\Sigma_i n_i \varepsilon_i = U_T \qquad (1.13)$$

The total number of ways of arranging N distinguishable atoms among the levels is $N!$. However, two similar atoms are mutually distinguishable only if they possess different energies. Thus permutation of the n_0 atoms among themselves do not produce distinguishable arrangements; similarly for all groups. The number of distinguishable arrangements W for this distribution is given by

$$W = \frac{N!}{n_0! \, n_1! \, n_2! \ldots n_i!} \qquad (1.14)$$

This is only one of many possible arrangements, consistent with eqns. (1.12) and (1.13). For example, a new arrangement with a different value of W could be produced by taking, say, three particles from any one level and replacing one in a level of energy twice that of the original and the other two in a level of half the energy. The most probable distribution is that set of $n_0, n_1, \ldots$, which makes W a maximum.†

In applying the condition for a maximum it is simpler in this case to consider the logarithm of W rather than W itself. Stirling's approximation ($\ln N! \simeq N \ln N - N$, for $N \gg 10$) may then be applied giving

$$\ln W = N \ln N - \Sigma_i n_i \ln n_i \qquad (1.15)$$

W and $\ln W$ are a maximum when the total derivative with respect to all the n_i's is zero. The partial derivatives are

$$\frac{\partial(\ln W)}{\partial n_i} = -(1 + \ln n_i)$$

and the condition for a turning value is

$$d(\ln W) = -\Sigma_i(1 + \ln n_i)dn_i = 0 \qquad (1.16)$$

The increments dn_i must all be consistent with eqns. (1.12) and (1.13). Thus, two other conditions are

$$\Sigma_i dn_i = 0 \qquad (1.17)$$

$$\Sigma_i \varepsilon_i dn_i = 0 \qquad (1.18)$$

Solution of eqns. (1.16) (1.17) (1.18) is obtained by the method of undetermined multipliers. Multiplying eqns. (1.17) and (1.18) by arbitrary constants α and β and adding the three equations,

$$\Sigma_i[(1 + \ln n_i) + \alpha + \beta \varepsilon_i]dn_i = 0 \qquad (1.19)$$

Since the dn_i's are never zero, eqn. (1.19) is satisfied only if all the coefficients are zero.

---

† It can be shown that when the number of particles in a system is large the value of W for the most probable arrangement, W_{max}, is so large compared with that for all other arrangements, that the latter may be neglected in comparison.

THE BASIS OF KINETIC THEORY

Therefore
$$(1 + \ln n_i + \alpha + \beta \varepsilon_i) = 0$$
which on rearrangement becomes
$$n_i = A e^{-\beta \varepsilon_i} \tag{1.20}$$
where $A = e^{-(\alpha+1)}$. The value of A is obtained by noting that
$$N = \Sigma_i n_i = A \Sigma_i e^{-\beta \varepsilon_i} \tag{1.21}$$
Therefore
$$n_i = \frac{N e^{-\beta \varepsilon_i}}{\Sigma_i e^{-\beta \varepsilon_i}} \tag{1.22}$$

β is determined as follows. From the Boltzmann equation and (1.15)
$$S = k(N \ln N - \Sigma_i n_i \ln n_i) \tag{1.23}$$
Putting eqn. (1.22) into logarithmic form gives
$$\ln n_i = \ln N - \beta \varepsilon_i - \ln \Sigma_i e^{-\beta \varepsilon_i} \tag{1.24}$$
Multiply each eqn. (1.24) by its own value of n_i; e.g.
$$n_0 \ln n_0 = n_0 \ln N - n_0 \beta \varepsilon_0 - n_0 \ln \Sigma_i e^{-\beta \varepsilon_i}$$
Adding all such equations
$$\Sigma_i n_i \ln n_i = N \ln N - \beta U_T - N \ln \Sigma_i e^{-\beta \varepsilon_i} \tag{1.25}$$
and combining (1.23) and (1.25) gives
$$S = k\beta U_T + kN \ln \Sigma_i e^{-\beta \varepsilon_i} \tag{1.26}$$
Therefore
$$\left(\frac{\partial S}{\partial U_r}\right)_V = k\beta \tag{1.27}$$
One of the Maxwell relations is
$$\left(\frac{\partial S}{\partial U}\right)_V = \frac{1}{T} \tag{1.28}$$
and so
$$\beta = 1/kT \tag{1.29}$$

Equation (1.22) becomes

$$\frac{n_i}{N} = \frac{e^{-\epsilon_i/kT}}{\Sigma_i e^{-\epsilon_i/kT}} \qquad (1.30)$$

In deriving eqn. (1.30) it was assumed for simplicity that all quantum states have different energies. However, quantum mechanics permits several microstates (quantum states) to have the same energy, i.e. to be degenerate. The number of states with the same energy is termed the degeneracy g of that state. For example, if two states, the j and the $(j+1)$, have the same energy ϵ_j, $g_j = 2$ and the number of particles in the combined level is twice that in the j level if it were not degenerate. By repeating the previous derivation whilst allowing all levels to be degenerate gives

$$\frac{n_i}{N} = \frac{g_i e^{-\epsilon_i/kT}}{\Sigma_i g_i e^{-\epsilon_i/kT}} \qquad (1.31)$$

Equation (1.31) describes the most probable distribution of thermal energy in a system of N particles and, according to the assumption of the method, is the one which will be realized in a system in equilibrium. This distribution is known as the *Maxwell–Boltzmann*, the *Boltzmann* or simply the *classical distribution law*: the fraction of particles in a system having an energy ϵ_i is proportional to the Boltzmann factor $e^{-\epsilon_i/kT}$. In deriving this result no restrictions were placed upon the type of energy. It applies to the total energy or to the component parts of the total (e.g. vibrational, translational, rotational, etc.) taken individually or in any combination.

Although some quantum mechanical principles were used in the above derivation, to be strictly accurate Maxwell–Boltzmann statistics apply only to a group of particles obeying the laws of classical dynamics. Two alternative distribution laws, the Einstein–Bose and the Fermi–Dirac, describe the behaviour of particles governed by quantum mechanics. The difference between quantum and classical statistics is negligible in nearly all transformation problems, except in a few special cases such as the behaviour of helium at low temperature.

THE BASIS OF KINETIC THEORY 19

The summed Boltzmann factor in the denominator of eqn. (1.31) is called the *partition function* Q, a most important function in statistical mechanics. Once Q is known all the thermodynamic functions of the system can be calculated and chemical equilibria can be expressed in terms of it in a simple way. Some important properties of partition functions are given in Appendix A.

1.9. The Number of Particles having Sufficient Thermal Energy to overcome a Potential Energy Barrier of Height U_A^0

Particles taking part in a reaction have to surmount the potential barrier U_A^0 (see Fig. 1.2). The number having sufficient thermal energy to enable them to do this may now be derived from eqn. (1.30). However, in carrying out the calculation it is necessary to be more specific concerning the type of energy. For example, the thermal energy of complex gas molecules consists mainly of translational and rotational energy of the molecules and vibrational energy of the component atoms but some of these forms of energy may not be relevant to the reaction under consideration.

In solids the atoms are constrained to oscillate about fixed sites and the thermal energy is vibrational energy only. In the Einstein model of a solid, to which most metals conform at temperatures at which reaction rates are reasonable, the atoms are assumed to vibrate completely independently in three dimensions. This is equivalent to saying that a system of N three-dimensional oscillators may be replaced by $3N$ particles vibrating independently in one dimension.

The energy levels of a simple harmonic oscillator are given by $(i + \tfrac{1}{2})\mathbf{h}\nu$ where $\mathbf{h}$ is Planck's constant, ν the frequency of oscillation and i the quantum number of the level. Hence $\varepsilon_0 = \tfrac{1}{2}\mathbf{h}\nu/kT$,

$$\varepsilon_1 = \frac{3\mathbf{h}\partial}{2kT}$$

and so on. The fact that the levels are evenly paced makes the problem comparatively simple. The partition function for the initial configuration is

$$Q_I = \sum_{i=0}^{i=\infty} e^{-(\frac{1}{2}+i)h\nu/kT} = e^{-\frac{1}{2}h\nu/kT} \sum_{i=0}^{i=\infty} e^{-ih\nu/kT} \quad (1.32)$$

Because the energy levels are all equally spaced U_A^0 may be equated to a fixed number of quanta, say n. That is,

$$U_A^0 = nh\nu \quad (1.33)$$

n is thus the quantum number for the level corresponding to U_A^0. From the distribution law the fraction of the total atoms having an energy equal to U_A^0, that is in the nth level, is

$$Q_I^{-1} e^{-(n+\frac{1}{2})h\nu/kT}$$

The fraction in the nth and higher levels is

$$f = Q_I^{-1}[e^{-(n+\frac{1}{2})h\nu/kT} + e^{-(n+\frac{3}{2})h\nu/kT} + e^{-(n+\frac{5}{2})h\nu/kT} \ldots]$$
$$= Q_I^{-1}[e^{-\frac{1}{2}h\nu/kT} \cdot e^{-nh\nu/kT}(1 + e^{-h\nu/kT} + e^{-2h\nu/kT} + \ldots)]$$
$$= Q_I^{-1} e^{-\frac{1}{2}h\nu/kT} \cdot e^{-nh\nu/kT} \cdot \sum_{i=0}^{i=\infty} e^{-ih\nu/kT} \quad (1.34)$$

and in view of equation (1.32) this reduces to

$$f = e^{-nh\nu/kT} \quad (1.35)$$

or
$$f = e^{-U_A^0/kT} \quad (1.36)$$

This result, whilst derived for a model likely to be of wide application for metals and alloys is by no means general. Its simplicity is a result of the fact that the energy levels of an oscillator are equally spaced. This is not so for other forms of energy. However, Hinshelwood[†] has shown that the same result applies for energy possessed in any two degrees of freedom, for example rotational and vibrational. For more complex energy distributions the expression for f includes the Boltzmann factor together with pre-exponential terms involving both U_A^0 and T. However, the variation of these terms with U_A^0 and T is negligible compared with the variation of the exponential and to a good approximation over a reasonable temperature range can be neglected.

The results of this and the preceding section can be generalized thus: in any system at constant T and V in equilibrium the fraction of the total number of particles having a thermal energy not less

---

† C. Hinshelwood, *Kinetics of Chemical Change*, O.U.P., 1940.

than a specified value U_A^0 measured relative to the zero point energy is proportional to $e^{-U_A^0/kT}$, the constant of proportionality being unity in many cases.

The units of U_A^0 are discussed in Appendix B.

1.10. The Rate of a Single Thermally Activated Process

Using the results in the preceding section it is possible to derive a general expression for the rate of a transformation which involves only one basic atomic process characterized by a unique activation energy U_A. It is assumed that U_A is independent of temperature (cf. (1.7)) so that $U_A = U_A^0$. The rate is given by the fraction of the total number of particles which reach the final configuration in unit time. It is obviously proportional to:

(a) The frequency with which particles " attempt " to transform. In solids this is equivalent to the vibration frequency v. In reactions in solutions or in the gas phase this frequency is the frequency of collisions between the reactant particles.

(b) The fraction of the particles in the initial equilibrium state having sufficient energy to surmount the potential barrier, given by $e^{-U_A^0/kT}$.

(c) The probability p that during the time a particle or particles have the requisite energy they satisfy the geometrical or other conditions necessary for the change. For example, for an atom to move into an adjacent vacant site it must be moving in the direction of the site during the time of activation; for a simple cubic lattice this would give $p = 1/6$. As more complicated processes are considered the geometrical conditions become more complex and the probability that they are satisfied decreases. For reactions involving the co-operative interaction of several particles p is very small.

Thus at constant temperature T, the rate of reaction is

$$\text{rate} = dy/dt = pve^{-U_A^0/kT} \tag{1.37}$$

At first sight it is difficult to reconcile this result with the idea developed in eqn. (1.6) that the free energy of activation G_A

determines reaction rates. However there is a close connection between entropy and probability as emphasized by the Boltzmann equation. In fact p is simply the ratio of the number of complexions associated with the transition configuration and that associated with the initial state, i.e.

$$p = W_A/W_I \tag{1.38}$$

and from eqn. (1.11) it follows that

$$p = e^{S_A/k} \tag{1.39}$$

Hence

$$dy/dt = v e^{S_A/k} \cdot e^{-U_A^0/kT} \tag{1.40}$$

Since $G_A = U_A - TS_A$ and $U_A = U_A^0$ eqn. (1.40) may be written

$$dy/dt = v e^{-G_A/kT} \tag{1.41}$$

This equation is the most succinct form of the result of this simple quasi-equilibrium theory of reaction rates.

It is common to combine v and the entropy term into one term, A, i.e.

$$dy/dt = A e^{-U_A/kT} \tag{1.42}$$

A is known as the *frequency factor*. The logarithmic form of (1.42) is

$$\ln(dy/dt) = \ln A - (U_A/kT) \tag{1.43}$$

Equation (1.41) or any of its various forms, is known as the *Arrhenius* equation.

If both A and U_A are independent of temperature a graph of ln (rate) against $1/T$ is linear, the gradient being $-U_A/k$ and the intercept on the rate axis $\ln A$. A vast number of physical and chemical reactions, both homogeneous and heterogeneous behave in this way, indicating that this model of a rate process is reasonable.

The magnitude of U_A relative to the thermal energy, which is of order kT, completely dominates the reaction rate. For example, if $U_A = 10,000$ cal/mole changing the temperature from 500 to 1000°K increases the rate by a factor of 150 approximately. Doubling U_A would decrease the rate at 500°K by a factor of 2×10^4 and at 1000°K by a factor of 150. The larger the value of U_A the more rapid is the variation of reaction rate with temperature. With the order of magnitude of U_A obtaining for most processes in metals, 10–70,000 cal/mole, a decrease of a few hundred degrees is enough virtually to stop a reaction which is taking place with moderate velocity at the higher temperature. This is the basis for quenching which is a process used to preserve at room temperature a phase which exists in equilibrium at high temperature but which would decompose during equilibrium cooling.

1.11. Further Remarks about the Theory

It is important to realize that this model of a rate process is a quasi-equilibrium model. In using eqn. (1.36) for the number of energically favoured particles it is assumed implicitly that the equilibrium distribution of thermal energy which this equation describes is not distorted by the continuous removal of some of the high energy particles. It is of interest to examine whether or not this is a reasonable assumption. In a typical process, $U_A \doteqdot 10,000$ cal/mole and for $T = 1000°K$ the probability that any one atom has a thermal energy not less than U_A is $e^{-U_A/RT} \simeq 1/150$. p is seldom greater than 1/10 and usually much less. Thus for every 1500 vibrations of a particle not more than 1 leads to its removal from the distribution whilst the remainder result in collisions by which the energy distribution is maintained. Thus, in most transformations the rate at which atoms leave the initial state is exceedingly small and the assumption of an undistorted energy distribution in the initial state is plausible.

A more subtle assumption implied in this derivation is more difficult to justify. U_A and $(G_A$ and $S_A)$ has a unique value only if the transition state has a unique value of internal energy dependent only upon the state. But according to thermodynamics those properties are single-valued state variables only when the system is in equilibrium. Hence the assumption that U_A^r is single valued is equivalent to assuming that the transition state is an equilibrium state.

1.12. A Simple Alternative Derivation

It is of interest to note that on the basis of the assumed equilibrium conditions it is possible to derive eqn. (1.42) in a simple alternative way. If the number of particles in the transition state per unit volume is c_A and in the initial state is c_I then it is possible to write down an equilibrium constant K^* for the equilibrium between the two configurations.

$$K^* = c_A/c_I \qquad (1.44)$$

K^* is given by the thermodynamic relation

$$K^* = e^{-G_A/kT} \qquad (1.45)$$

The reaction rate must be proportional to C_A and so

$$dy/dt = \text{rate} = \text{const } C_I \cdot e^{-G_A/kT} \qquad (1.46)$$

or

$$dy/dt = A e^{-U_A/kT} \qquad (1.47)$$

In eqn. (1.47) the entropy component of G_A has been taken into the frequency factor A.

Although the method emphasizes the assumption that the activated state is a quasi-equilibrium state it fails to reveal the physical significance of the model and particularly of U_A and S_A, and for this reason the former approach is preferred.

In the following section a further general theory is developed which also leads to closely similar results.

1.13. The Transition State Theory

The transition state theory, otherwise known as the theory of absolute reaction rates or the activated complex theory was largely developed by Eyring, Wigner, Polanyi and others primarily with chemical reactions in mind. But it is quite general and it has been applied widely with considerable success to many rate processes in metals. It is also a quasi-equilibrium theory, assuming that an equilibrium is established and maintained between particles in the initial state and those in the transition state (the activated complexes). The initial and final states are separated by a potential energy barrier, the height of which, allowing for the thermal energies of the reactants and products, is, as before, U_A.

The potential energy curve along the reaction path is that shown in Fig. 1.2. The transition state is here defined as all the configurations corresponding to an arbitrary distance $\frac{1}{2}l$ on either side of the maximum of the potential energy curve, measured along the reaction path. The problem of evaluating l does not arise because it does not appear in the final expression. The transition state is regarded as being physically real with unique values of the thermodynamic functions.

On the basis of this model the rate is given by the concentration of activated complexes c_A (i.e. the number per unit volume) times the average frequency with which they cross the saddle point. If the average velocity along the reaction path over the length l is $\bar{v}$ then the frequency of crossing is $\bar{v}/l$ and so

$$\text{rate} = c_A \bar{v}/l \qquad (1.48)$$

The problem is to express c_A, $\bar{v}$ and l in terms of measurable or calculable properties of the system.

Because the activated complexes are assumed to be in equilibrium the Maxwell–Boltzmann equation for the distribution of velocities may be applied giving:

$$\bar{v} = \left(\frac{kT}{2\pi m^*}\right)^{\frac{1}{2}}{}^\dagger \qquad (1.49)$$

where m^* is the effective mass of a complex.

c_A may be expressed in terms of the equilibrium constant K^* for the equilibrium between reactants and complexes. Let the reaction be represented by the general equation

$$xX + yY \rightarrow \text{complexes} \rightarrow cC + dD \qquad (1.50)$$

---

† Proof of this will be found in any standard text on Physical Chemistry.

Then

$$K^* = \frac{c_A}{c_X^x c_Y^y} \qquad (1.51)$$

Combining eqns. (1.48) (1.49) and (1.51) gives

$$\text{rate} = \left(\frac{K^*}{l}\right)\left(\frac{kT}{2\pi m^*}\right)^{\frac{1}{2}} c_X^x \cdot c_Y^y \qquad (1.52)$$

The equation giving the relationship between the reaction rate and the concentration of reactants is known as a *rate equation*. Equation (1.52) is of this type. The general form is

$$\text{rate} = k \cdot f(c)$$

where $f(c)$ is any function of the initial concentration and k is called the *rate constant*. The experimental determination of k is discussed in Chapter 2.

Thus the rate constant k is given by

$$k = \left(\frac{K^*}{l}\right)\left(\frac{kT}{2\pi m^*}\right)^{\frac{1}{2}} \qquad (1.53)$$

The next step is to express K^* in terms of the partition functions of the reactants and complexes. Using the result in Appendix A.

$$K^* = \frac{Q^*}{Q_X^x Q_Y^y} e^{-U_A^0/kT} \qquad (1.54)$$

The complexes are, however, rather special in that movement in one direction—across the barrier—leads to decomposition. If l is chosen small enough, the potential energy will be virtually constant over that length of the reaction path and movement along the path within this length defining the transition state is effectively translation in a constant potential field. The partition function for one degree of translation freedom within a length l is

$$\frac{l}{h}(2\pi m^* kT)^{\frac{1}{2}}$$

h is Planck's constant. This degree of freedom may be factorized out of the complete partition function for the activated state Q^*

$$Q^* = Q_2^*(2\pi m^* kT)^{\frac{1}{2}} l/h \qquad (1.55)$$

THE BASIS OF KINETIC THEORY

Q_2^* is the partial partition function for the remaining two degrees of freedom. Combining eqns. (1.53) (1.54) and (1.55) gives

$$k = \frac{1}{l}\frac{Q_2^*}{Q_X^x Q_Y^y}\left(\frac{kT}{2\pi m^*}\right)^{\frac{1}{2}}(2\pi m^* kT)^{\frac{1}{2}}\frac{l}{h}e^{-U_A^0/kT}$$

$$= \frac{kT}{h}\frac{Q_2^*}{Q_X^x Q_Y^y}e^{-U_A^0/kT} \quad (1.56)$$

The term $(Q_2^*/(Q_Y^y Q_Y^x))\,e^{-U_A^0/kT}$ has the form of an equilibrium constant (Appendix A).

Define

$$K_2^* = \frac{Q_2^*}{Q_X^x Q_Y^y}e^{-U_A^0/kT} \quad (1.57)$$

then

$$k = (kT/h)K_2^* \quad (1.58)$$

Equation (1.58) is the most generalized form of the result of the transition state theory. kT/h is a universal constant for a given temperature; it is the frequency with which a complex crosses the barrier. At room temperature its value is about 6×10^{12}.

The disadvantage with this expression is that K_2^* is not a true equilibrium constant because it is derived from a ratio of partition functions with different numbers of degrees of freedom. The justification for regarding it as a true equilibrium constant is that the value of l may be chosen to make the translational partition function unity; in which case $Q^* = Q_2^*$.

On the assumption that K_2^* obeys the same equations as a true equilibrium constant the thermodynamic relation

$$G_A = -kT \ln K_2^* \quad (1.59)$$

may be applied. G_A is the free energy per atom absorbed in the formation of the complexes.

Therefore

$$k = \frac{kT}{h}e^{-G_A/kT} \quad (1.60)$$

or

$$k = \left(\frac{kT}{h}\right) e^{S_A/k} \cdot e^{-H_A/kT} \qquad (1.61)$$

where S_A and H_A are the entropy and enthalpy of activation. The difference between changes in enthalpy and internal energy is negligible in condensed systems so that eqn. (1.61) becomes

$$k = \left(\frac{kT}{h}\right) e^{S_A/k} \cdot e^{-U_A/kT} \qquad (1.62)$$

It is to be noticed that the transition state theory leads to a result very similar to that of the simple theory, the main difference being the inclusion of a temperature term outside the exponential. However, the conversion from K_2^* to G_A is not beyond question due to the incompatibility of the partition functions in eqn. (1.56). Zener† has shown how this difficulty can be overcome, for the restricted case of a process involving the motion of single atoms. The partition functions for the initial particle may also be factorized into two components, one for one-dimensional oscillation along the line to the transition state $Q_{x.1}$, and one for the other two degrees of freedom Q_{x2}. For a single atom process Q_X is factorized into

$$Q_X = Q_{X1} \cdot Q_{X2} \qquad (1.63)$$

the suffix denoting the number of degrees of freedom.

At high temperature the partition function for a particle oscillating in one dimension with frequency v is‡

$$Q_1 = \frac{kT}{hv} \qquad (1.64)$$

and so eqn. (1.56) becomes

$$k = \frac{kT}{h} \frac{Q_2^*}{(kT/hv)Q_{x.2}} e^{-U_A^0/kT} \qquad (1.65)$$

$$= v \left(\frac{Q_2^*}{Q_{x2}}\right) e^{-U_A^0/kT} \qquad (1.66)$$

---

† C. Zener, *Thermodynamics in Physical Metallurgy*, Am. Soc. Metals, 1950.
‡ See Appendix A.

THE BASIS OF KINETIC THEORY 29

The partition functions are now compatible and may properly be replaced by free energy expressions

$$k = v\, e^{S_A/k} . e^{-U_A/kT} \quad (1.67)$$

This expression is identical to that derived in section 1.10.

In cases other than the simple one treated by Zener, the theory is best applied by eqn. (1.56). However, in practice the partition function difficulties have often been ignored and eqn. (1.61) has been applied directly.

1.14. The Derivation of U_A and S_A

The calculation of the rate of a single activated reaction requires the calculation of the activation energy and the frequency factor for the basic atomic process.

The derivation of U_A for a specific atomic process starts with the derivation of U_A^1, the increase in potential energy of an atom or group of atoms as it, or they, move from the initial potential energy through to the saddle point configuration. Calculation of this function requires a knowledge of the forces binding the atoms in the initial state and the mutual interaction forces as atoms move from their normal sites. Electron theory is not yet sufficiently advanced to permit completely rigorous solutions for even the simplest processes. In the most generalized form the derivation of U_A^1 consists of calculating the potential energy surface for all conceivable reaction paths and finding the most favourable one. The problem is formidable. Limited success has been achieved for a few very simple gaseous chemical reactions using semi-empirical methods. Considerable simplification is possible when the saddle-point configuration is self-evident. Examples are the migration of point defects through crystalline solids for which the point of maximum potential energy is obvious from lattice geometry. But even in these cases, considerable approximations are required to facilitate the mathematical treatment and so calculation of U_A^1 has been attempted for only a few extremely simple processes. The calculation of U_A^0 and U_A from U_A^1 requires a knowledge of the thermal energy levels in both initial and activated states. These are given by the solutions of Schödinger's equation for the appropriate conditions. So little is known about the activated state that it is generally impossible to carry out these calculations.

Thus at the present time very few calculated activation energies are available and these are very approximate and relate to relatively simple atomic processes. Fortunately, the empirical activation energy E_A for a macroscopic process can be readily derived from experimental observations by determining the variation of ln (rate) with $1/T$. Details of this and other methods are discussed in Chapter 2. When there is no doubt that the reaction observed is the result of only one atomic process E_A may be identified with U_A for the elementary step. Activation energies determined in this way can then be used to analyse the kinetics of more complex processes.

It cannot be emphasized too strongly that almost all reactions involve a number of different basic microscopic processes each with its own characteristic activation energy, and in these cases it is not possible to attach any significance to the experimental activation energy determined from the slope of an Arrhenius graph. The value observed will depend upon the relative contributions and interdependence of the various steps. To stress that, in general, empirical activation energies differ from the theoretical activation energy the former is given the special symbol E_A. In particular cases E_A may achieve fundamental value in the light of a detailed model of the reaction.

The calculation of S_A requires a detailed knowledge of the activated configuration and since this can never be isolated for experimental study it is possible only to make reasonable guesses about its nature and then use the comparison with experimental results as a check on the accuracy of the assumptions. For some simple atomic processes reasonably successful attempts have been made by the use of semi-empirical methods. For example, Zener† related the change in the vibrational entropy accompanying

---

† See Chapter 3.

the movement of interstitial solutes through a b.c.c. lattice with the temperature dependence of the elastic constants. It is also unfortunate that considerable uncertainty attends the determination of an empirical activation entropy. The intercept of the Arrhenius graph gives a value of ln A. But in practice A contains several unknown quantities arising from the technique used to study the change—see (2.8).

Thus it is seen that kinetic theory in its present form must undergo extensive development before transformation or reaction velocities can be derived *ab initio*.

1.15. Reactions Involving Several Atomic Processes

Previous considerations were confined to homogeneous reactions which involve only one basic atomic process. Heterogeneous reactions involve at least two, and often more, consecutive steps, each of which has its own characteristic activation energy. The frequently made generalization that the overall rate of reaction is that of the slowest step is true only if the rate of the slowest step is very much slower than the others.† If this condition is satisfied, otherwise complicated kinetics reduce to the simpler kinetics of a single activated process and the activation energy E_A of the overall reaction is that of the slow step. If these conditions are not satisfied then the overall rate is a complicated function of the rates of all the steps. A simple mechanical analogy shown in Fig. 1.3 will serve to illustrate the point. Vessel A is maintained at a constant level of water H_A from an infinite reservoir. It discharges through an outlet O_A into a second vessel B which has an outlet O_B. If O_B is very large compared to O_A then no water will accumulate in B and the flow of water from the system is simply the flow through O_A under the constant head H_A. The other limiting case is when $O_B \ll O_A$. Then water fills up in B until it reaches the level of that in A. Flow from A through O_A maintains it constant. The rate of discharge through O_B is then quite independent of O_A.

---

† If the various component steps are completely independent the overall rate is always that of the slowest step, irrespective of the relative speeds of the components. This situation seldom, if ever, arises in real processes.

At intermediate sizes of O_B, the level in B reaches a steady state value H_B at which the flow through O_A under a head $(H_A - H_B)$ is equal to that through O_B under a head H_B. The flow from the system may be calculated in terms of either orifice but H_B must enter into the calculation and since this is related to the relative sizes of O_A and O_B, both appear in the final expression for the flow.

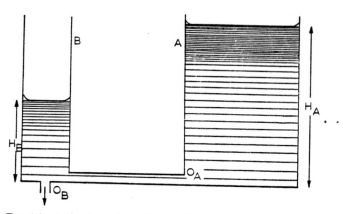

Fig. 1.3. A simple analogy of a reaction involving two consecutive steps. Water is maintained in a tank A at a constant level H_A and passes into B through O_A. It discharges from B through O_B. H_B depends upon the relative sizes of O_A and O_B. If $O_B \ll O_A$, $H_B \simeq H_A$ and the flow from O_B is independent of O_A. For $O_B \gg O_A$ $H_B \simeq 0$ and the flow is independent of O_B. For intermediate cases the flow depends upon both O_A and O_B.

A physical equivalent situation exists during the growth of a precipitate from a solution. Diffusion through the solution replaces O_A; crystallization of atoms onto the new lattice at the interface is the second stage. The composition in the solution adjacent to the interface depends upon the relative rates of these two processes. In the limiting case of very fast crystallization the growth is described as diffusion controlled and the growth kinetics are diffusion kinetics and the activation energy that of diffusion. The other limiting case is when the growth is controlled by the interface reaction, the diffusion being very rapid. In other circumstances the rates of both are important and the growth rate is a function of both. The empirical activation energy for growth then has no unambiguous physical significance.

If simultaneous processes are involved in a composite reaction the overall rate is the sum of the different processes. At low temperatures that having the lowest activation energy contributes most. At higher temperature the one with highest U_A predominates. In these reactions strictly Arrhenius type temperature dependence can only exist if the activation energies of the various simultaneous processes are equal. In other cases the Arrhenius graph is a curve, the slope at low temperature tending to that characteristic of the lowest activation energy step; and at high temperature the limiting slope is that of the highest activation energy. An example is referred to on p. 70. In practice it is often found that reactions which are undoubtedly of this type give linear Arrhenius graphs within the experimental error provided the temperature range is not too great. The empirical activation energy derived therefrom is some kind of a weighted mean of those of the contributing processes.

Minor deviations from Arrhenius behaviour are to be expected for even single atomic processes because in general neither the pre-exponential term nor U_A are independent of temperature. These minor deviations produce only slight curvature in the Arrhenius graph and in practice it would require extremely precise experimental techniques to observe the effects.

1.16. The Principle of Maximum Reaction Velocity

If there are a number of different ways for a metastable configuration to transform the one actually observed is that which gives the maximum rate of decrease of free energy, i.e. that which has maximum reaction rate. Two cases should be distinguished; first, when different reaction paths lead to the same final state, and second when different products are produced. The first case is equivalent to that already referred to in (1.6) where it was pointed out that for energetic reasons the path used is that having the lowest activation energy. This is also the path of maximum velocity.

The case of different products is slightly more complicated. Although the most stable structure is the most likely product because the driving force is a maximum, it often happens that less stable configurations are produced preferentially. As a result of random thermal fluctuations the atoms attempt, quite impartially, to surmount all barriers leading to a reduction in free energy. The preponderant product is that produced at the fastest rate. Because reaction rate is largely determined by the value of U_A the product produced most rapidly is that associated with minimum activation energy. Furthermore, since a small change in U_A produces a large change in $e^{-U_A/kT}$, a metastable state whose activation energy for formation is smaller than that for the stable configuration, forms in preference. In the multistage processes observed in practice each stage has a greater activation energy than its fore-runner. Two metallurgical examples will illustrate the point.

(a) In the example of the precipitation hardening of aluminium —4 per cent copper alloys discussed earlier—the activation energy for the formation of [G.P.I.] zones is less than for the other possible products. Consequently it forms in preference thereby achieving maximum rate of decrease of free energy. When this stage is complete the configuration of next highest activation energy is formed and so on, until the most stable state is produced.

(b) The decomposition of austenite may occur by several alternative processes, one of which involves diffusion and produces the stable state of ferrite plus cementite. Another is diffusionless producing the metastable constituent martensite. The details of these transformations will be discussed in greater detail in later chapters. It is sufficient here to note that below a certain temperature the rate of the former mode decreases rapidly as the temperature decreases whereas the rate of the diffusionless reaction is far less affected by temperature. Consequently at high temperature the formation of ferrite and cementite is favoured. At low temperature the rate of the diffusion process is sensibly zero and the diffusionless process occurs because, although the free energy decrease which attends it is not as large as that which would accompany the diffusion process if it were possible, the fact that it has a finite rate results in a finite rate of decrease of free energy.

Further Reading for Chapter 1

Thermodynamics and Statistical Mechanics

DARKEN, L. S. and GURRY, R. W., *Physical Chemistry of Metals*, McGraw-Hill, 1953.
SLATER, J. C., *Introduction to Chemical Physics*, McGraw-Hill, 1939.
RUSHBROOKE, G. S., *Introduction to Statistical Mechanics*, Clarendon Press, Oxford, 1949.
SWALIN, R. A., *Thermodynamics of Solids*, Wiley, 1964.

Kinetic Theory

HINSHELWOOD, C. N., *Kinetics of Chemical Change*, Clarendon Press, Oxford, 1940.
GLASSTONE, S., LAIDLER, K., EYRING, H., *Theory of Rate Processes*, McGraw-Hill, 1939.

General

Thermodynamics in Physical Metallurgy, Am. Soc. Metals, 1950.

CHAPTER 2

Empirical Kinetics

2.1. Alternative Definitions of Reaction Velocity

The reaction velocity was defined in Chapter 1 as the rate of change of the fraction transformed y. It is equally meaningful and sometimes more convenient to express the reaction rate as the rate of change of the concentration of one of the reactants or products. Any reaction may be represented by the general equation

$$xX + yY = cC + dD \tag{2.1}$$

The velocity at any time t may be specified by any of the following: $-dc_X/dt$; $-dc_Y/dt$; dc_C/dt; dc_D/dt. c_X, c_Y, c_C and c_D are the concentrations of the components at time t.

The fraction transformed $y(t)$ may be defined in terms of any component, e.g.

$$y(t) = \frac{c_X(0) - c_X(t)}{c_X(0) - c_X(\infty)} \tag{2.2}$$

where $c_X(0)$, $c_X(t)$ and $c_X(\infty)$ are the concentrations of X, initially, at time t and after completion of the reaction, respectively.

Hence the relation between dy/dt and dc_X/dt is

$$dy/dt = -\frac{1}{c_X(0) - c_X(\infty)} \cdot dc_X/dt \tag{2.3}$$

In general these various expressions for the reaction rate have different numerical values. However, they are all proportional to each other. For example, x moles of X disappear for every y moles of Y and thus

$$dc_X/dt = \frac{x}{y} \cdot \frac{dc_Y}{dt}$$

EMPIRICAL KINETICS 37

The proportionality factors are constants for fixed starting conditions. In practice, it is seldom that the numerical values have any significance; only the variation of velocity with temperature, pressure, composition, microstructure and other experimental variables is important. Consequently, the proportionality constants may be ignored and any of these definitions of velocity used.

2.2. The Methods of Measuring Reaction Velocities

In essence the determination of reaction rate consists of the determination of the concentration, or other quantity unit, of a phase as a function of time, yielding dc/dt directly. This direct approach is used only occasionally for transformations in solid metals and alloys. Quantitative metallography and X-ray techniques can be used for analytical purposes but in general they are of poor accuracy, are slow and tedious and require the examination of either (a) a series of samples reacted for various times, all other experimental conditions being held constant or (b) one sample, the reaction being interrupted by some means whilst the analysis is carried out. Either method demands that the reaction can be stopped at any required instant, the usual way being to rapidly decrease the temperature. The difficulty with (a) is to ensure consistency from sample to sample; the objection to (b) is that the interruptions (usually involving quenching and then re-heating) may change the kinetics.

The alternative approach is to observe the change of some property that is a function of the concentration of one of the phases. The measurement of changes in linear dimensions or volume (dilatometry), or in electrical resistivity or magnetic properties are most common. Changes in mechanical properties, particularly hardness, are also used. Measurement of the evolution of the heat of reaction has fundamental appeal on thermodynamic grounds but the practical difficulties associated with the smallness of the quantities involved prevent its widespread use. The advantages of utilizing the change in some physical property are rapidity and ease of automatic continuous recording making it possible to study a reaction continuously in one sample without disturbance of the temperature or other conditions. The disadvantage is that absolute values of concentrations are not obtained and the relationship between the particular property and constitution is rarely, if ever, available. In principle, it is possible to carry out a calibration but this is usually not considered worthwhile in view of the difficulties associated with quantitative analysis in solids. Instead, it is generally assumed that a linear relationship exists between the value of the property observed and the fraction transformed or concentration of one of the components. On the basis of this assumption the reaction rate is equal to the rate of change of the physical property. It is important to realize that this assumption is often seriously in error because properties such as hardness and resistivity are sensitive to other factors as well as constitution; particle size and lattice coherency being but two of the possible interfering side effects.

More detailed discussions of the experimental methods used in kinetic investigations are given in the references at the end of this chapter.

2.3. The Rate Equation and the Rate Constant

At fixed values of the temperature and other experimental variables the rate of a reaction is a function of the time due to the fact that reaction rates depend upon the concentration of the reactants and these change continuously as the reaction proceeds. It is thus inconvenient to use the numerical values of velocity in kinetic discussions because it is necessary to refer each value to the appropriate fraction of transformation. This difficulty is overcome by expressing the results in terms of a *rate equation*. A rate equation expresses the functional dependence of the rate on concentration or fraction transformed. The general form is

$$dc_x/dt = k_c f(c) \qquad (2.4a)$$

or in terms of y

$$dy/dt = k_y f(y) \qquad (2.4b)$$

in which $f(c)$ and $f(y)$ are any functions of c and y respectively and k_c and k_y are constants known as the *rate or velocity constant* or the *specific reaction rate*. $f(c)$ and k_c or $f(y)$ and k_y are determined from experimental data. They are a compact and useful means of describing the reaction under the particular experimental conditions. Often $f(c)$ or $f(y)$ do not change over the variations of the temperature and other external conditions used in practice in which case the complete kinetic characteristics may be described by the variations in the rate constant.

Experimentally it is y or c or some property related to them that is measured, and not dy/dt. Consequently eqn. (2.4) must be put into a different form. Separation of the variables and integration gives

$$g(y) = k_y t \qquad (2.5a)$$

or

$$g(c) = k_c t \qquad (2.5b)$$

where $g(y)$ and $g(c)$ are further functions of y and c. Equations (2.5a) and (2.5b) are generalized forms of *integrated rate equations*. The functions $g(y)$ and $g(c)$ are determined empirically by finding the functions which describe the variations of y (or c) with time. In principle this may be done by the standard procedures of data analysis but it is unlikely that a completely arbitrary function would be useful for theoretical purposes. From this point of view it is better to compare the data with functions derived for specific processes because this often gives valuable clues as to the detailed model capable of explaining the results.

A number of rate equations are available which have proved to be widely applicable as empirical relationships. The usual practice is to test the data against each of these to find which, if any, represent the observations satisfactorily. If all the common equations fail then an attempt is made to derive alternative forms on the basis of assumed models of the process. Some of the most useful of the standard equations are presented in the subsequent sections.

Once the integrated rate equation is established the rate constant k_y or k_c and the function $f(y)$ and $f(c)$ may be determined. It is to be noted that k does not have a unique value for a given set of data but depends upon the function chosen to represent the data. It sometimes happens that two empirical rate equations fit the observations equally well, in which case it is possible to get two different values for k.

The dimensions of k_y are $(\text{time})^{-1}$; k_c is $(\text{time})^{-1}$ times concentration raised to some power depending upon the form of the function $f(c)$. Occasionally, integrated rate equations are formulated in such a way that the constant does not have the dimensions of $(\text{time})^{-1}$. An example appears on p. 46. These are not true rate constants and cannot be compared with constants having the correct dimensions.

2.4. Rate Equations for Homogeneous Reactions—First and Second Order Reactions

Reactions in the gaseous phase or in liquid solutions clearly depend upon the frequency with which the requisite number of the various kinds of molecules come into contact during the course of random motion. The collision frequency is proportional to the probability of the required number of particles being in the same small volume element at the same instant. The probability of one completely free molecule of a particular kind being in a small volume element is proportional to the number of that kind per unit volume, i.e. to the concentration c. The probability of finding x molecules of X and y molecules of Y in the same volume element is proportional to $c_X^x \cdot c_Y^y$. Thus for those reactions involving collisions between freely moving particles it is plausible to suppose the rate of reaction to be proportional to the concentration of the reactants raised to various powers. In this case the rate equation takes the general form

$$- dc_X/dt = k_c c_X^x \cdot c_Y^y \tag{2.6}$$

For this restricted type of expression there is defined the *order* of reaction o equal to the sum of the powers of the concentrations appearing in the empirical rate equation. The power of a component is the order with respect to that component. The rate eqn. (2.6) is of order $(x + y)$ and of order x with respect to X.

For a *first order* reaction the rate equation becomes
$$- dc/dt = k_c c \tag{2.7}$$
Separating the variables and integrating using the condition $c = c_0$ at $t = 0$ gives
$$\ln(c/c_0) = -k_c t \tag{2.8}$$
In terms of y this becomes
$$\ln(1 - y) = -k_y t \tag{2.9}$$
The applicability of this result is tested by plotting the logarithm of the experimentally measured values of c/c_0 against the time. If the reaction is first order the graph is linear of slope $-k$. If common logarithms are used the gradient is $-k/2\cdot3$. Another useful method of testing is by measurement of the fractional reaction times. The time to half transformation is $1/k\,(\ln 2)$ and is independent of the concentration for a first order reaction. The second half-life in a single run is $-1/k \ln(\frac{1}{4}c_0/\frac{1}{2}c_0)$ which is $1/k\,(\ln 2)$, i.e. the successive fractional times in a single run are equal.

The value of k may be determined using either of these two procedures. A third method is also useful. Equations (2.8) and (2.9) show that $1/k$ is equal to the time at which the concentration of reactants falls to $1/e \simeq 36\cdot79$ per cent of its initial value. $1/k$ has the dimensions of time and is often referred to as the "relaxation time" for the particular process.

For a *second order* change involving only a single reactant the rate equation is
$$- dc/dt = k_c c^2 \tag{2.10}$$
which integrates to
$$\frac{1}{c} - \frac{1}{c_0} = k_c t \tag{2.11}$$
using the same boundary condition as in the first order case.

In terms of the fraction transformed this becomes
$$y/(1 - y) = k_y t \tag{2.12}$$
with $k_y = k_c/c_0$.

For a second order reaction a graph of $1/c$ or $(y/1 - y)$ against t is a straight line of slope k_c or k_y respectively. The relaxation time is equal to the half-life for a second order change. Successive half-lives are in the ratio of $1:2:4$.

First and second order reactions are compared in Fig. 2.1.

On the basis of simple collision ideas it was thought that a one-to-one correspondence existed between the order of a rate equation and the number of particles participating in the basic

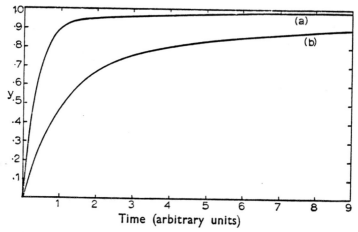

FIG. 2.1. Isothermal reaction curves for (a) a first order reaction and (b) a second order reaction. y is the fraction transferred. The rate constant is the same for both curves.

reaction event. If this were so then the order should always be a small, positive whole number. Although this is usual, it is by no means universal and fractional or even negative orders are observed. Furthermore, in general the empirical order is often different from that to be expected from the stoichiometric equation. Evidently, this interpretation of the kinetic order is untenable. It ignores the fact that seemingly simple processes may involve a complicated mechanism involving several basic steps. Simple processes give a simple rate equation the order of which has fundamental significance but the converse is not necessarily true. Complex processes often yield misleadingly simple kinetics. For example, if a reaction occurs by a sequence of steps, one of which is very much slower than the others, the observed kinetics will be essentially that of the slow reaction. In general, therefore, the order has purely empirical significance as a parameter describing a restricted type of rate equation.

In solids the movement of the atoms is constrained to a fixed lattice and simple collision ideas can hardly be expected to apply. However, the concept has proved useful as a means of classifying the kinetics of a limited number of solid state reactions. The annealing out of supersaturated point defects produced by irradiation, cold work or quenching is one such case. This is a particular example of a complex process giving apparently simple kinetics. The rate of disappearance of excess point defects is dependent upon the type and initial distribution of the defects, their rates of diffusion and the nature and distribution of sinks. In certain circumstances the resulting time dependence is to a good approximation that of a first or second order process emphazing that empirical parameters achieve fundamental significance only when analysed in terms of a specific physical model of the process.

2.5. The Rate Equation for an Autocatalytic Reaction

When one of the products of a reaction is a catalyst for the reaction, the reaction velocity is a function of both the concentrations of the reactant and product. The phenomenon is called *autocatalysis*. It can occur in both homogeneous and heterogeneous reactions. The situation is encountered often in solid state transformations, because such transformations are usually accelerated by stress. The product of transformation has, in general, a different specific volume from that of the parent phase and the volume change is accommodated by elastic and plastic deformation of parent and product. The level of the associated internal stress is dependent upon the amount of product phase present and the elastic modulii of the two phases. At low temperatures, metals are comparatively rigid and the transformation induced stress correspondingly high. These considerations are particularly relevant to martensitic reactions which are very stress sensitive and usually occur at low temperatures.

The simplest autocatalysis case, kinetically, is when the process is first order with respect to the reactant. Then the rate is given by

$$dy/dt = ky(1 - y) \tag{2.13}$$

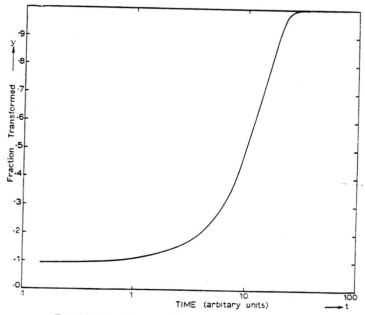

FIG. 2.2. Reaction curve for an autocatalytic reaction.

where k is the rate constant. Integration gives

$$\ln y/(1 - y) = kt + I \tag{2.14}$$

I is an arbitary constant. The usual boundary condition, $y = 0$ at $t = 0$ makes eqn. (2.14) indeterminate. Thus it must be presumed that either a small fraction y_0 of the system is already transformed before transformation proper begins, in which case

$I = \ln y_0/(1 - y_0)$; or alternatively, that the equation is not applicable at early stages. The transformation curve is S-shaped as shown in Fig. 2.2, the velocity increasing from zero to a maximum at $y = \frac{1}{2}$ and then decreasing to zero. Transformation curves of this general type are designated *sigmoidal*. The autocatalytic sigmoidal curve differs from others by being symmetrical about $y = \frac{1}{2}$ when time is plotted on a linear scale. The applicability of eqn. (2.14) is tested by plotting $\log y/(1 - y)$ against t which, if the equation fits, is a straight line of gradient equal to the rate constant $k/2 \cdot 3$.

2.6. Empirical Rate Equations for Heterogeneous Reactions

Heterogeneous systems consist of a mixture of phases and reaction in such a system occurs by the growth of one or more phases at the expense of others. In general, each phase is not found as one single entity but as a dispersion of smaller domains. Transformation involves the formation of new domains of product referred to as *nucleation* and the advancement of the phase boundaries termed *growth*. The rate of such a transformation depends upon the rate of nucleation and of growth of individual domains and the effect of mutual interference of neighbouring domains either through direct impingement or by long range competition for solute atoms. The kinetics of growth of phases in the solid state is discussed in Chapter 6, of nucleation in Chapter 5 and of the overall kinetics of some idealized models in Chapter 7. However, the kinetics of many heterogeneous reactions are too complex to be treated analytically and it is necessary to resort to empirical solutions.

It is found empirically that an equation of the general form

$$dy/dt = k^n t^{n-1}(1 - y) \tag{2.15}$$

describes the isothermal kinetics of a wide variety of reactions in metals. Furthermore, many of the theoretical kinetic equations derived for simple processes may be reduced either actually or approximately to this same form with specific values of k and n; k has the dimension (time)$^{-1}$ but is not a true rate constant because it is defined by an equation involving both y and t—see eqn. (2.3).

The reaction rate is small to begin with, increases to a maximum and then decreases to zero due to the impingement effects. The factor $(1 - y)$ may be regarded as an allowance for the retardation in reaction rate due to impingement. Assuming k and n to be true constants independent of y (and thus of t) at constant temperature allows eqn. (2.15) to be integrated, thus

$$\ln \frac{1}{1-y} = (kt)^n \qquad (2.16)$$

in which term $1/n$ has been taken into the constant. Equation (2.16) yields a sigmoidal rate curve. An equivalent form is

$$y = 1 - e^{-(kt)^n} \qquad (2.17)$$

Occasionally eqn. (2.15) is written

$$dy/dt = kt^{n-1}(1-y)$$

the corresponding form of (2.17) being

$$y = 1 - e^{-kt^n}$$

In this form k has the dimensions of time^{-n}. Activation energies derived from the temperature dependence of this constant cannot be compared directly with values derived from constants having dimensions of time^{-1}. The problem was discussed by Zener[†] who showed how the difficulty could be overcome. However, it is preferable to avoid the difficulty by using the form in eqn. (2.17).

Equation (2.17) is referred to as the Johnson–Mehl equation, being a generalized form of a particular equation, with $n = 4$, derived by these authors for the special case of the formation of pearlite from austenite.[‡] The time exponent n and the rate constant k are useful empirical parameters providing a concise

---

[†] C. Zener, *Trans. A.S.M.* **41**, 1057 (1949).
[‡] W. A. Johnson and R. F. Mehl, *Trans. A.I.M.E.* **135**, 416 (1939).

EMPIRICAL KINETICS   47

description of isothermal reaction kinetics when (2.17) is obeyed. k may take any positive value. In practice, if $k > 1 \text{ sec}^{-1}$ the reaction is too fast to be followed experimentally. n also may take any positive value; 0·5 to 2·5 is the most common range but values up to 5 or 6 are occasionally found.

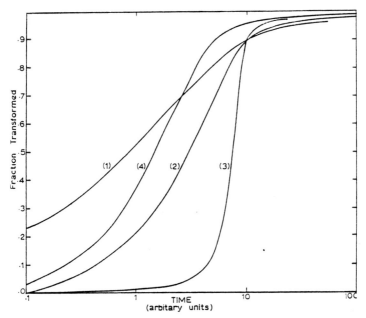

FIG. 2.3. Reaction curves conforming to the Johnson–Mehl equation. Curves (1), (2) and (3) have the same value of k and $n = \frac{1}{2}$, $n = 1$ and $n = 4$ respectively. Curve (4) has $n = 1$ and k half the value of the other curves.

Some rate curves which conform to eqn. (2.17) are shown in Fig. 2.3. This illustrates that when time is plotted on a logarithmic scale the shape of the curve is determined only by the index n; the constant k fixes the position on the time axis. Many transformations in metals have an incubation or induction period during which no detectable transformation occurs. In these cases

t in eqn. (2.17) should, strictly, be measured from the end of the incubation period. However, in practice it is very difficult to establish a reaction " start " time precisely and it is more meaningful and convenient to measure t from one common zero, usually the time the specimen attains the reaction temperature. Apart from slight distortions of the curve initially this causes negligible error.

It is often observed that the reaction curves for a particular process over a range of values of an experimental variable are of the same shape and may be brought into identity merely by lateral shift along the log time axis. The curves are then said to be *isokinetic*. If they obey the Johnson–Mehl equation they have the same value of n.

Converting eqn. (2.16) into common logarithms,

$$\log \frac{1}{1-y} = \frac{1}{2 \cdot 3}(kt)^n \qquad (2.18)$$

and taking logarithms again

$$\log \log \frac{1}{1-y} = n \log t + n \log k - \log 2 \cdot 3 \qquad (2.19)$$

Hence if a reaction conforms to the Johnson–Mehl equation a graph of $\log \log [1/(1-y)]$ versus $\log t$ is linear. Typical plots are shown in Fig. 2.4. The value of n is obtained from the slope and k from the intercept. However, it is better to obtain k directly from the data or the transformation curve. Rearrangement of eqn. (2.17) shows that $1/k$ equals the time at which

$$y = \frac{e-1}{e} = 0 \cdot 6321$$

An alternative method of determining the time exponent of a sigmoidal rate curve is occasionally used.† In essence it is based upon a re-definition of the equation for dy/dt. As noted in connection with eqn. (2.15), this is not a true rate equation in the

---

† M. Hillert, *Acta Met.* **7**, 653 (1959).

EMPIRICAL KINETICS

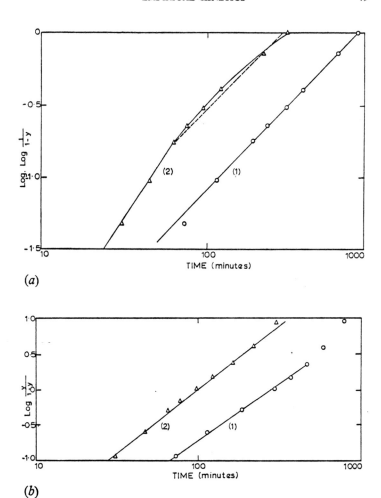

(a)

(b)

Fig. 2.4. (a) Typical graphs of $\log \log (1/(1 - y))$ against $\log t$. Curve (1) conforms to the Johnson–Mehl equation with $n = 1 \cdot 2$. Curve (2) deviates from the Johnson–Mehl equation for $y > \frac{1}{2}$. The broken line is a possible representation of the data between $y = \frac{1}{2}$ and $y = 1$ with $n \simeq 0 \cdot 8$. (b) The same data as in (a) replotted as $\log (y/(1 - y))$ against $\log t$.

sense of section 2.3 nor is the constant k a conventional rate constant because the equation for dy/dt involves both y and t. Since y is a function of t it is, in principle, possible to replace t by the appropriate function in y. The problem is to find this function. Dropping the impingement factor from eqn. (2.15) gives

$$dy/dt \propto t^{n-1}$$

and so

$$y \propto t^n \qquad (2.20)$$

whence

$$dy/dt \propto y^{(n-1)/n} \qquad (2.21)$$

It is thus plausible to assume that a relation of the form of eqn. (2.21) applies. Reintroducing the impingement factor $(1 - y)$ yields an alternative rate equation

$$dy/dt = k_m(1 - y)y^{(m-1)/m} \qquad (2.22)$$

This defines another rate constant k_m and exponent m different from n. It is important to realize that although m is of the same order as n it is not equal to it. Equations (2.21)–(2.22) should not be held to imply such an equality. Their purpose is merely to suggest a plausible assumption for the dependence of dy/dt on y. n is an exponent of t, to be determined empirically from eqn. (2.17) derived from arbitrary assumptions concerning the relation between dy/dt and t. m is an exponent of y arising out of the arbitrary assumptions about dy/dt as a function of y. Equation (2.22) is the more satisfactory in that it is a true rate equation so that k_m is a true rate constant and, further, y is a more natural independent variable than t with which to express reaction rates.

The big advantage of eqn. (2.22) over (2.15) is that the exponent m may be determined without the necessity of integration and therefore without the necessity of assuming k_m and m to be constants.

Taking logarithms gives

$$\ln(dy/dt) = \ln k_m + \ln(1 - y) + \left(\frac{m-1}{m}\right)\ln y \qquad (2.23)$$

EMPIRICAL KINETICS 51

Differentiating with respect to $\ln y$

$$\frac{d(\ln dy/dt)}{d(\ln y)} = \frac{d(\ln k_m)}{d(\ln y)} - \frac{y}{1-y} + \frac{m-1}{m} \quad (2.24)$$

The procedure is to determine the gradient of the y–t graph—the transformation curve—at each of a series of values of y. Plot the result as $\ln(dy/dt)$ against $\ln y$ and determine the gradient of the graph—the left-hand side of eqn. (2.24)—at each value of y required. The only unknown in eqn. (2.24), apart from m itself, is then $d(\ln k_m)/d(\ln y)$. There is no way of determining k_m directly as a function of y, but if it is thought to be independent of y then the first term in eqn. (2.24) vanishes. Alternatively, an indirect method, which does not require k_m to be independent of y, may be used. In view of the considerations in Chapter 1 it is possible to write

$$k_m = A e^{-E_A/RT} \quad (2.25)$$

where T is the absolute temperature, R the gas constant and A and E_A are the frequency factor and activation energy respectively. Using the method given in section (2.8c), it is possible to determine empirically the variation of A and E_A with y, quite independently of any assumptions concerning the form of the rate equation. These results permit the evaluation of $d(\ln k_m)/(d(\ln y))$ and thus of m as a function of y.

The error in this determination of m is the accumulated errors in the measurement of y as a function of t and E_A and A as a function of y and of the gradient determinations. In fact, an observed variation of m would have to be quite large to be significant. Whether the extra work required to find m instead of n is justified can only be decided in individual cases.

The Austin–Rickett equation. When the Johnson–Mehl graph of $\log \log 1/(1-y)$ against $\log t$ shows a pronounced negative curvature as in curve (2) in Fig. 2.4a better agreement is frequently obtained by replacing $(1-y)$ by $(1-y)^2$ in eqn. (2.15). In this case the rate equation becomes

$$dy/dt = (1-y)^2 k_A^n \cdot t^{n_A - 1} \quad (2.26)$$

which, assuming k_A and n_A to be constant, integrates to

$$\frac{y}{1-y} = (k_A t)^{n_A} \qquad (2.27)$$

with the term in n_A taken into the constant. This integrated rate equation was first used by Austin and Ricketts† to analyse the kinetics of austenite decomposition. Equation (2.27) defines another exponent n_A and rate constant k_A, which are different from either n or m defined by eqns. (2.17) and (2.22). n_A is determined from a graph of log $y/(1-y)$ against log t which is linear of slope n_A if eqn. (2.27) applies.

Curve (2) in Fig. 2.4 is drawn from transformation data which may be equally well described by either (1) the Johnson–Mehl equation using two values of n (Fig. 2.4a or (2) the Austin–Rickett equation using one value of n_A Fig. 2.4b).

The alternative method of writing the Johnson–Mehl equation (eqn. 2.22) may be adopted for the Austin–Rickett equation to define another exponent m_A. The equation corresponding to eqn. (2.22) is

$$dy/dt = k_{1M}(1-y)^2 y^{(m_A - 1)/m_A} \qquad (2.28)$$

where k_{1M} is a new rate constant. Taking logarithms and differentiating with respect to ln y gives

$$\frac{d(\ln dy/dt)}{d(\ln y)} = \frac{d(\ln k_{1M})}{d(\ln y)} - \frac{2y}{1-y} + \frac{m_A - 1}{m_A} \qquad (2.29)$$

which permits the evaluation of m_A as a function of y, using the same steps as described for the determination of m from eqn. (2.22).

2.7. The Empirical Activation Energy and Frequency Factor

With a few notable exceptions reaction rates increase rapidly with increasing temperature. Furthermore, provided that the temperature range is not too great the temperature dependence

---

† J. B. Austin and R. L. Ricketts, *Trans. A.I.M.E.* **135**, 396 (1939).

EMPIRICAL KINETICS 53

of the rates of most reactions obeys an Arrhenius type equation, i.e. a linear relation exists between the logarithm of the rate constant k and the reciprocal of the absolute temperature. This applies equally to homogeneous and heterogeneous reactions, simple or complex. In these circumstances it is always possible to define an empirical activation energy E_A and frequency factor A_A by the equation

$$k = A_A e^{-E_A/kT} \qquad (2.30)$$

It is worth re-emphasizing that it is only in the case of singly activated processes that E_A and A_A may be identified with the activation energy and frequency factor for the basic atomic event. Then, comparison of E_A and A_A with theoretical activation energies and entropies is a powerful means of elucidating the basic process of a reaction. In all other cases caution must be exercised in attaching fundamental significance to E_A and A_A.

2.8. The Determination of E_A and A_A

The general rate equation is

$$dy/dt = kf(y) \qquad (2.31a)$$

or
$$-dc/dt = kf(c) \qquad (2.31b)$$

The common logarithmic form of the Arrhenius equation for k is

$$\log k = \frac{\log A_A}{2 \cdot 3} - \frac{E_A}{2 \cdot 3 k} \cdot \left(\frac{1}{T}\right) \qquad (2.32)$$

E_A and A_A may be obtained from these relations in a number of ways.

(a) *The rate constant method.* Values of k are determined at various temperatures, all other experimental conditions being held constant. If E_A and A_A are constant a graph of $\log k$ against $1/T$ is a straight line. The gradient of this graph is $-E_A/2 \cdot 3 k$. The intercept on the $\log k$ axis gives the value of $(\log A_A/2 \cdot 3)$.

Although this method is the most direct it is not the most satisfactory. Its main attraction is that it gives A_A absolutely. One disadvantage arises because the determination of k from a set of isothermal measurements of y or c as a function of t depends upon the functions $f(y)$ or $f(c)$ selected empirically. This is evident from consideration of the procedures used to establish k from the various rate equations in sections 2.4–2.6. Thus, when this method is used it is essential to specify clearly the rate equation used because this partly defines the activation energy. A second disadvantage is that it is impossible to ascertain if E_A varies during transformation. Such a variation is not expected for simple reactions involving a single atomic process, the rate of which is described by eqns. (2.31a) and (2.31b) with k a true constant. However, the complex macroscopic processes which occur in alloys often involve several simultaneous or consecutive atomic events. At any instant the value of k for the process is the weighted mean of the individual constants. If the contribution made by the individual events changes as transformation proceeds the value of k varies with the progress of the transformation.† Such a change is manifested as a variation of E_A and/or the frequency factor A_A. This method of determining E_A yields an average value only.

Any variation of E_A or A_A with temperature is revealed as a curvature of the Arrhenius plot. The value of E_A at a particular temperature is obtained from the gradient at that temperature. For a reasonably narrow temperature range the curvature, if any, is likely to be so small as to justify taking an average value.

(b) *The time to a given fraction method.* If the value of E_A is to be meaningful it is desirable to make its determination independent of the empirical functions $f(y)$ and $f(c)$. Since y and t are functionally related it is possible to choose t instead of y as the dependent variable. Equation (2.31a) may then be written

$$dt = k^{-1} f^{-1}(y) dy \qquad (2.33)$$

The time, t_Y, required for a specified fraction $y = Y$ to transform is

$$t_Y = k^{-1} \int_{y=0}^{y=Y} f^{-1}(y) dy$$

---

† For example, in a nucleation and growth reaction it is extremely unlikely that nucleation will proceed for more than a fraction of the total transformation—see Chapter 5.

The reaction is studied at a series of temperatures, all other experimental variables being maintained constant and the time, t_Y, to the chosen value of y measured at each temperature. Provided that the function $f(y)$ does not vary over the temperature range studied, the integral has a constant numerical value. Hence

$$t_Y \propto k^{-1} \qquad (2.34)$$

$$t_Y \propto A_A^{-1} e^{E_A/kT} \qquad (2.35)$$

and $\qquad \ln t_Y = \text{const} - \ln A_A + E_A/k \left(\frac{1}{T}\right) \qquad (2.36)$

A graph of $\log t_Y$ against $1/T$ is linear if E_A and A_A are independent of T, the slope being $E_A/2\cdot 3\text{k}$. The procedure may be repeated for various values of y to reveal any variation of E_A with y. If such a variation exists then the value of E_A at $y = Y$ is some mean value from the start of reaction to $y = Y$. In this connection it is of interest to note that E_A is independent of y only if the reaction is isokinetic (see section 2.6). Considering two isokinetic curves for temperatures T_1 and T_2 the difference between $\log t_{T_1}$ and $\log t_{T_2}$, $\Delta \log t$, must be constant for all values of y and is the relative movement of the two curves required to produce identity. $\Delta(1/T)$ is, of course, constant and so E_A, which is equal to $\text{k}[(\Delta \log t)/\Delta(1/T)]$ is also constant. If the curves are not isokinetic, $\Delta/\log t$ must vary with y and so must E_A.

It is not possible to determine A_A by this method because as is seen from eqn. (2.36), the intercept on the $\log t_Y$ axis contains an unknown constant.

For illustration consider a reaction which is studied by measurement of the rate of change of the electrical resistance Φ. The value of Φ depends upon the resistivity ρ and the geometry of the specimen. In turn ρ varies with y, with the temperature of measurement θ, which may or may not be the same as the reaction temperature T and the composition. If the investigation is conducted on one specimen or a set of identical specimens and the treatment before reaction at each value of T ensures that the initial composition is always the same, then the change for a fractional transformation y is

$$\Delta\rho(\theta,y) = y\left(\frac{d\rho}{dy}\right) \tag{2.37}$$

Thus
$$\Delta\Phi(\theta,y) = X \cdot y\left(\frac{d\rho}{dy}\right) \tag{2.38}$$

where X is the factor which converts ρ to Φ (in this example length/cross-sectional area). It is a constant if the specimens are identical. Hence, for a fixed fractional transformation Y

$$\Delta\Phi_Y(\theta) = X \cdot Y \cdot (d\rho/dy) \tag{2.39}$$

When all measurements are made at one fixed temperature θ, $(d\rho/dy)$ is constant and

$$\Delta\Phi_Y = \text{const } Y \tag{2.40}$$

The activation energy is determined by noting the time $t_{\Delta\Phi}$ to a given change $\Delta\Phi$ as a function of reaction temperature T and plotting $\log(t_{\Delta\Phi})$ against $1/T$. However, it is not possible to associate this value of E_A with a particular value of y until $X(d\rho/dy)$ is determined. This is usually done by noting the values $\Phi(\theta)$ for $y = 0$ and $y = 1$ and assuming a linear relationship. If the actual value of Y is of no interest there is no necessity to study a complete reaction at any temperature, a consideration which may be of practical importance if the reaction is slow.

When the measurements of Φ are made at the reaction temperature then eqn. (2.40) only applies if $d\rho/dy$ does not vary with temperature. Frequently, however, $d\rho/dy$ does vary and it is necessary to study the complete change at each temperature and by assuming a linear relationship between Φ and y derive t for $y = Y$ directly.

(c) *The change of rate method.* If $(dy/dt)_{T,Y}$ is the rate of reaction at a fixed value of $y = Y$ and at a temperature T_1 and $(dy/dt)_{T,Y}$ is the rate at the same value of y but a different temperature T_2, provided that the function $f(y)$ in eqn. (2.31) is constant in the temperature range T_1–T_2 and that all other

EMPIRICAL KINETICS 57

experimental conditions remain the same dy/dt is proportional to k. That is

$$\frac{(dy/dt)_{T_1 Y}}{(dy/dt)_{T_2 Y}} = \frac{k_{T_1}}{k_{T_2}} \qquad (2.41)$$

Generalizing,

$$(dy/dt)_{Y,T} = k_T \qquad (2.42)$$

where T is any temperature in the range for which $f(y)$ is constant, and k_T is the rate constant at the temperature T. It follows that

$$\log(dy/dt)_Y = \text{const} + \log A_A - \frac{E_A}{2 \cdot 3k}\left(\frac{1}{T}\right) \qquad (2.43)$$

This result is applied in two ways. Firstly, the complete reaction is studied at each of a series of temperatures, with all other conditions held constant and the gradient of the reaction curves determined graphically at various values of y. A family of graphs of $\log(dy/dt)$ against $1/T$ is plotted, one for each value of y; the gradient of this reciprocal temperature plot yields the value of E_A at the corresponding value of y. Thus the variation of E_A with y may be determined unambiguously. A_A cannot be determined because of the unknown constant in eqn. (2.43).

The second method is to partly react a specimen at one temperature T_1 and then suddenly to change the temperature to T_2. The gradient of the reaction curve is measured just before the change of temperature—$(dy/dt)_{T_1,Y}$—and immediately after — $(dy/dt)_{T_2,Y}$. At the instant of change y, and hence $f(y)$, is the same for both temperatures. Thus

$$\log\left(\frac{dy}{dt}\right)_{T_1,Y} - \log\left(\frac{dy}{dt}\right)_{T_2,Y} = \frac{E_A}{2 \cdot 3k}\left(\frac{1}{T_1} - \frac{1}{T_2}\right) \qquad (2.44)$$

assuming that the frequency factor varies negligibly with temperature. To be strictly correct the physical property measurements should be carried out at a fixed reference temperature θ to ensure that $(d\rho/dy)_{T_1} = (d\rho/dy)_{T_2}$. However, the range $T_1 - T_2$ is usually small, about 10°C is typical, and it is reasonable to neglect any change of $(d\rho/dy)$ over this range. The advantage of this

c

method is that only one specimen is required and it does not require the study of a complete reaction at any temperature. Against this must be set the increased uncertainty in the result due to the use of only two temperatures and the fact that E_A is obtained at one unknown value of y only. The accuracy of both methods is limited by the accuracy with which the gradients can be determined graphically.

(d) *Other methods.* Other methods of determining E_A are used occasionally.

(i) E_A can be determined from the gradient of a graph of $\ln[(1/y)(dy/dt)]$ against $1/T$ for fixed values of y. Since y is a constant in this procedure

$$d\left(\ln \frac{1}{y}\frac{dy}{dt}\right) \text{ is the same as } \frac{1}{y}d\left(\ln \frac{dy}{dt}\right)$$

and thus the method is identical with method c.

(ii) During the annealing out of point defects it is frequently found that recovery occurs in a series of well defined stages each confined to a narrow temperature range. Thus when the rate of annealing is plotted as a function of temperature, the rate is small and constant except over those ranges which are characterized by a peak. Each peak is associated with the disappearance of one type of defect, and each occurs when a particular activated process achieves a sensible speed. By assuming a rate law for each process it is possible to relate the temperature of the peak with an activation energy. In this way activation energies can be determined from isochronal curves. However, this method is restricted to processes which occur over a very sharply defined temperature range and it requires the assumption of a rate law. It is not sufficiently general to warrant discussion in detail.

2.9. Time–Temperature–Transformation Diagrams

It is evident from the preceding sections that temperature is the most important variable in any kinetic study. A useful method of displaying the effect of temperature on a particular reaction in any one alloy is by means of a Time–Temperature–Transformation diagram, usually referred to as a *TTT* diagram. The times to the start and end of a reaction and to various fractions transformed

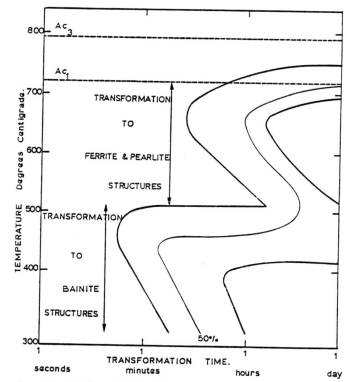

FIG. 2.5. Isothermal Time–Temperature–Transformation diagram for the decomposition of austenite in a low alloy steel (B.S. En 100). Two transformations are possible each producing a *C*-curve.

are plotted usually on a logarithmic scale against the temperature. It is difficult to determine the start and finish times of a reaction accurately and so it is usual to take the time to 5 and 95 per cent transformation instead.

A common type of *TTT* curve for transformations in metals and alloys involving a change from a high temperature phase to a low temperature phase during isothermal annealing at temperatures below the equilibrium transition temperature takes the form of a *C*-curve. The reaction requires a long incubation period and, once started, proceeds slowly both at high temperature near to the transition temperature and at low temperatures; the minimum incubation time and maximum velocity is realized at some intermediate temperature corresponding to the nose of the *C*-curve. The physical basis for *C*-curve kinetics is discussed in (5.3). Figure 2.5 is an example in which the parent is able to transform in two ways, each producing a *C*-curve. In many such cases the *C*-curves overlap and cannot be resolved as in Fig. 2.5.

It is more satisfactory to plot *TTT* diagrams in terms of the reciprocal of the absolute temperature instead of temperature. For if any section is linear the activation energy may then be derived directly from the graph as in method B in (2.8) and any change in E_A with fraction transformed is revealed as non-parallel graphs for the various values of *y*.

Further Reading for Chapter 2

SEYBOLT, A. U. and BURKE, J. E., *Procedures in Experimental Metallurgy*, Wiley, 1954.

CHALMERS, B. and QUARRELL, A. G., *Physical Examination of Metals*, Arnold, 1960.

Modern Research Techniques in Physical M tallargy, Am. Soc. Metals, 1953.

DIENES, G. J. and VINEYARD, G. H., *Radiation Effects in Solids*, Interscience, 1957.

CAHN, J. W., and HAGEL, W. C., *Theory of the Pearlite Reaction in Decomposition of Austerite by Diffusional Processes* Interscience, 1962.

HINSHELWOOD, C. N., *Kinetics of Chemical Charge*, Clarendon Press, Oxford, 1940.

CHAPTER 3

Diffusion in Metals

3.1. Introduction

The thermally activated transport of atoms through matter is known as diffusion. Flow through a bulk, polycrystalline metal or alloy may occur by several different diffusion processes (a) movement of atoms through the lattice, referred to as volume or lattice diffusion, (b) diffusion along the surface, (c) diffusion along grain boundaries, and (d) diffusion along dislocations and other plane and line defects. Each process has its characteristic activation energy and therefore to obtain meaningful information about any one process, it is advantageous to design the diffusion experiment so that the contribution of the others to the total flow is negligible. By using well-annealed single crystals or large grained polycrystals it is possible to reduce the volume of material in grain boundaries or dislocations to a very small fraction of the total and it is then justifiable to assume that the observed flow is due solely to lattice diffusion. Lattice diffusion is also the simplest to treat theoretically because of the greater atomic regularity. The other diffusion processes are more difficult to study both experimentally and theoretically and will not be considered.

3.2. The Diffusion Coefficient

Diffusion is superficially similar to heat conduction. Consequently Fick proposed that the diffusive flow of an atomic species is proportional to the composition gradient, in the same way as heat flow is proportional to the temperature gradient.

For unidirectional flow in the x-direction, this gives for each species

$$J_x = -D_x \frac{\partial c}{\partial x} \qquad (3.1)$$

c is the concentration of the species considered at the point x expressed as gram-atoms or number of atoms per unit volume and J_x, the flux in the x-direction, is the number of gram-atoms or atoms flowing through unit area in unit time. The rate constant D_x defined by eqn. (3.1) is the *diffusion coefficient* or *diffusivity* of the particular species in the x-direction. D has the units of length^2 time^{-1}. The negative sign indicates that diffusion occurs from high to low concentration.

Although Fick's first law (eqn. (3.1)) was proposed on empirical grounds it is also possible to derive it on the basis of a simple, idealized model. Suppose diffusion is the result of the migration of single atoms through the crystal by a series of random and unrelated jumps between adjoining sites. Then if there are two planes in the crystal A and B and there are more atoms of a particular kind on A than on B, a net flow occurs from A to B because, statistically, more atoms jump from A to B than from B to A. Let A and B be of unit area, normal to the x-axis, and separated by a distance, b_x, equal to the jump distance along the x-axis. If the concentration at A is c_A the number of atoms on plane A is $c_A.b_x$; similarly, there are $c_B.b_x$ atoms on plane B. Let the mean frequency with which an atom leaves a site irrespective of direction averaged over a large number of jumps be v_J. If the probability that any one jump is in the $+x$-direction is p_x then the number of jumps made per unit time by one atom in the $+x$-direction is $p_x.v_J$. The number of atoms leaving A and arriving at B in unit time is $(p_x.v_J.c_A.b_x)$ and the number making the reverse journey is $(p_x.v_J.c_B.b_x)$. The net gain of atoms at B, given by J_x is

$$J_x = p_x v_J b_x (c_A - c_B) \qquad (3.2)$$

Provided that $(c_A - c_B)$ is sufficiently small that c is a continuous

DIFFUSION IN METALS 63

and single-valued function of x, $(c_A - c_B)/b_x$ may be replaced by the negative of the derivative of c with respect to x

Hence
$$J_x = -p_x \cdot v_J b_x^2 \cdot \partial c/\partial x \qquad (3.3)$$
which is Fick's first law with D_x identified as
$$D_x = p_x \cdot v_J b_x^2 \qquad (3.4)$$

Values of p_x and b_x are readily derived. For example, cubic lattices are isotropic and all six orthogonal directions are equally likely, i.e. $p_x = \frac{1}{6}$. For f.c.c. metals the distance between first nearest neighbours is $a/\sqrt{2}$, a being the lattice parameter. Hence for f.c.c. crystals eqn. (3.4) becomes
$$D_x = D_y = D_z = (\tfrac{1}{12}) \cdot v_J \cdot a^2 \qquad (3.4a)$$
For b.c.c.
$$b = (\sqrt{3}a/2)$$
and
$$D_x = D_y = D_z = (\tfrac{1}{8}) v_J a^2 \qquad (3.4b)$$
For simple cubic $b = a$ and thus
$$D_x = D_y = D_z = (\tfrac{1}{6}) v_J \cdot a^2 \qquad (3.4c)$$

At this stage v_J cannot be defined more fully because it depends on the mechanism by which the atoms move through the lattice.

Fick's first law can be applied experimentally only if a steady state exists in which the concentration at every point is invariant, i.e. $(\partial c/\partial t) = 0$ for all x. The non-stationary flow equation is derived from eqn. (3.1) by considering the rate of increase of the number of atoms $(\partial c/\partial t)$ in a unit volume of specimen. This is equal to the difference between the flux into and that out of the volume. If the specimen is of unit cross-sectional area, unit volume is defined by two planes a unit distance apart. The flux across one plane is J_x and across the other $(J_x + 1 \cdot \partial J/\partial x)$, the difference being $-(\partial J/\partial x)$.

Hence
$$\frac{\partial c_x}{\partial t} = -\frac{\partial J_x}{\partial x} = \frac{\partial}{\partial x}\left(D_x \frac{\partial c}{\partial x}\right) \qquad (3.5)$$

Equation (3.5) is Fick's second law. If D is independent of concentration eqn. (3.5) becomes

$$\frac{\partial c_x}{\partial t} = D_x \frac{\partial^2 c}{\partial x^2} \tag{3.6}$$

Generalization of Fick's laws for three-dimensional diffusion is readily made

$$\frac{\partial c}{\partial t} = \frac{\partial}{\partial x}\left(D_x \frac{\partial c}{\partial x}\right) + \frac{\partial}{\partial y}\left(D_y \frac{\partial c}{\partial y}\right) + \frac{\partial}{\partial z}\left(D_z \frac{\partial c}{\partial z}\right) \tag{3.7}$$

The purely mathematical problem of deriving solutions of Fick's second law for various boundary conditions in isotropic and anisotropic materials is fully covered in texts on the mathematics of diffusion or heat conduction and will not be considered. The principles of diffusion theory can be described adequately in terms of unidirectional diffusion in isotropic (e.g. cubic) systems in which $D_x = D_y = D_z$. In view of this, the subscript x will be dropped from future equations.

In discussions of diffusion it is necessary to distinguish between several different diffusion coefficients. In a chemically pure metal atoms move through the lattice under the influence of thermal agitation. This movement is referred to as *self-diffusion*. It can be studied by observing the penetration into a crystal of a radioactive isotope of the same metal. Except in the most sophisticated analysis the slight difference between the masses of the isotopes may be neglected and the rate of diffusion of the radioactive tracer is that of self-diffusion. The coefficient characterizing this diffusion is called the *self-diffusion* or *tracer self-diffusion coefficient*. Self-diffusion is the simplest case to study for the purpose of relating the kinetics of atom movement to basic atomic properties because the electronic structure and the interatomic forces are better understood in pure metals than in alloys; and, secondly, since every atom is identical and has the same environment, the simple statistical concept of diffusion leading to eqn. (3.4) applies.

In a homogeneous alloy, the self-diffusion of each component can also be measured by a tracer method to give the *tracer diffusion coefficients* of the components of a given alloy of specified composition. In general the coefficients for the various components are not equal and vary with composition.

Diffusion in alloys that are not chemically homogeneous is termed *chemical diffusion*, and is manifested by the gradual disappearance of composition gradients. Although technically the most important type of diffusion, this is by far the most difficult to understand in fundamental terms. The variation of the thermodynamic parameters of the system with composition means that the simple statistical approach is totally inadequate. Completely random jumping is possible only when every configuration of atoms has the same energy as every other, which is true only in pure elements or in ideal solutions, in which the internal energy is independent of the atomic distribution. In non-ideal solutions the random thermally activated jumps are weighted by the tendency of the atoms of the system to take up low energy configurations. For example, in a solution of A and B atoms if the bonding between A and B atoms is stronger (i.e. are of lower energy) than the mean of $A-A$ and $B-B$ pairs, A atoms tend to jump towards B atoms rather than towards A atoms to lower the energy of the system. Further, the jump frequency is also related to the average strength of the bonds between an atom and its neighbours and this also varies with composition.

The rate of diffusion in a chemical gradient is a complex function of the tracer diffusion coefficients of the components, the composition and the thermodynamic properties of the system; Fick's equations are not applicable. Notwithstanding these remarks, it is practically convenient, if only because a great deal of effort has gone into solving Fick's equations for a variety of boundary conditions, to retain its form and to define a *chemical diffusion coefficient* $\tilde{D}$ in terms of eqn. (3.5):

$$\frac{\partial c}{\partial t} = \frac{\partial}{\partial x}\left[\tilde{D}\, \frac{\partial c}{\partial x}\right] \qquad (3.8)$$

where $(\partial c/\partial t)$ is the rate of change of composition at a distance x from the origin. D is thus a measure of the rate of interpenetration of two metals at a given alloy composition. It varies with composition. Whilst the physical significance of D may be obscure, it is of great utility and many applications of diffusion data to practical problems are made in terms of eqn. (3.8).

Equation (3.8) is not in itself a complete definition of D because it is necessary to specify the origin of coordinates from which x is to be measured. This selection is by no means as obvious as it may appear and further discussion is postponed until section 3.4.

3.3. The Correlation Factor

Even for self-diffusion in a pure metal it is an over-simplification to assume that jumping is completely random. Certain diffusion mechanisms give rise to *correlated motion* in which the direction of any diffusion jump is influenced by the direction of the previous jump. For example, anticipating later discussion, if movement is by an atom exchanging places with a vacancy, there is a greater probability that an atom will return to its initial site than move on to another because the probability of there being a vacancy with which it can exchange is greater for its original site than for any other. Correlation between successive jumps may be incorporated into the statistical definition of D in eqn. (3.4) by the introduction of a correlation factor f, such that

$$D = p \cdot v_J \cdot b^2 \cdot f \qquad (3.9)$$

f is the ratio of the square of the displacement produced by a given number of correlated jumps to the square of the displacement produced by the same number of uncorrelated jumps. The value of f for self-diffusion depends on lattice geometry and the diffusion mechanism and has been calculated for a number of particularly important cases. Its value which lies in the range 0–1 is found to be independent of temperature for self-diffusion, but it may have a marked temperature dependence for diffusion in alloys.[†]

[†] K. Compaan and Y. Haven, *Trans. Faraday Soc.* **52**, 786 (1956); A. D. Leclaire, *Phil. Mag.* **7**, 141 (1962)

3.4. Experimental Studies of Diffusion

The main object of a diffusion experiment is the determination of a diffusion coefficient and its variation with various factors. Self-diffusion coefficients are the most important fundamentally and can be measured with high accuracy by tracer methods. A common arrangement consists of depositing a very thin layer of radioactive isotope onto a plane surface of a well-annealed single crystal or large-grained polycrystal, and allowing diffusion to occur by annealing for a known time, in suitable environments to prevent contamination. After the anneal, the concentration of the tracer as a function of the distance from the original interface is determined by sectioning and counting or by photographic methods. Provided that diffusion does not reach the end of the specimen the solution of eqn. (3.6) for this arrangement is

$$c(x,t) = M \exp - \left(\frac{x^2}{4Dt}\right) \qquad (3.10)$$

where $c(x,t)$ is the concentration at a distance x from the original surface after annealing for a time t, and M is a constant related to the quantity of tracer deposited per unit area of the surface.

A graph of ln c as a function of x^2 is linear of slope $-1(4Dt)$. since t is known D can be calculated from the measured gradient. Absolute values of c are not required. Only some property which is proportional to c, such as the number of counts recorded by a Geiger counter, need be measured because the multiplication of c by a constant in eqn. (3.10) does not alter the gradient of the graph, only its position.

The usual method of studying chemical diffusion is to butt-weld specimens of the two metals or alloys across a plane interface and determine the composition–distance curve after a diffusion anneal of known time. The curve is seldom symmetrical about the original interface indicating that more atoms leave one side of the weld than are replaced by diffusion from the other, i.e. that the two species diffuse at different rates. Equation (3.8) must be solved to obtain values of $\bar{D}$. An approximate method derived by

Boltzmann and first applied by Matano is commonly used. It consists of replacing the two independent variables x and t by one variable λ given by

$$\lambda = x/t^{1/2} \tag{3.11}$$

Substitution into eqn. (3.8) gives

$$\lambda/2 \cdot \frac{dc}{d\lambda} = -\frac{d}{d\lambda}\left(\tilde{D}\frac{dc}{d\lambda}\right) \tag{3.12}$$

and rearranging

$$\tilde{D}(c) = -\tfrac{1}{2}\left(\frac{d\lambda}{dc}\right)\int_{c}^{c_0}\lambda \cdot dc \tag{3.13}$$

in which $\tilde{D}(c)$ is the value of $\tilde{D}$ at the concentration c, and c_0 is the initial concentration on one side of the couple (Fig. 3.1). Since t, the time of the anneal, is a constant for a given specimen, the variable λ may be replaced by x using eqn. (3.11) to give

$$\tilde{D}(c) = -\frac{1}{2t}\left(\frac{dx}{dc}\right)_c\int_{c}^{c_0}x\,dc \tag{3.14}$$

The problem of defining an origin from which to measure x now arises. This was referred to in connection with the definition of $\tilde{D}$. The mass flow given by Fick's laws is the difference between the currents in opposite directions. This can be seen from the derivation of eqn. (3.3). Hence $x = 0$ is that plane across which the net flow in opposite directions is zero. When $\tilde{D}$ is independent of c this is the original interface. In the case of variable $\tilde{D}$ this is not so as shown by the asymmetry of the penetration curve. A typical curve is shown in Fig. (3.1) in which the side marked 1 has suffered a net loss of atoms.

For the purpose of eqn. (3.14) it is necessary to choose as origin that plane across which the net flow is zero, i.e. the plane, $I_M I_M^1$, which makes the two shaded areas in Fig. (3.1) equal. This plane is called the Matano interface. Mathematically it is defined by the relation

$$\int_0^{c_0} x\,dc = 0 \tag{3.15}$$

DIFFUSION IN METALS

This definition of the zero of coordinates completes the definition of $\bar{D}$—cf. section 3.2. In practice it is determined by trial and error either by counting the squares on the graph paper or by planimeter measurements. An important point to note is that concentrations must be expressed in atomic units and not weight per cent. The integral in eqn. (3.14), equal to the area $A\ B\ C\ I_M\ I_0$ is estimated graphically, as also is the gradient $(dx/dc)_c$ at the

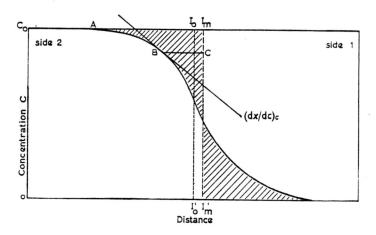

FIG. 3.1. The Matano–Boltzmann graphical method of deriving chemical diffusion coefficients from a composition–distance curve. $I_0 I_0^1$ is the original interface and $I_m I_m^1$ the Matano interface.

value of c chosen. The procedure is repeated at various values of c to reveal the composition dependence of $\bar{D}$.

The initial interface between the two halves of a chemical diffusion couple can be identified by the use of inert markers. These may be fine wires or foils of ceramics or high melting point metals which do not dissolve significantly during the time of the experiment. It has been found that, in a wide variety of systems, such markers are displaced relative to the ends of the specimen, the direction of the movement being towards the side rich in the lower melting point component. This movement, known as the *Kirkendall effect*, is an important factor in considerations of the mechanism of diffusion (cf. section 3.7). It arises when the rate of flow of the components of an alloy relative to the lattice are unequal. In a binary alloy the atoms of the low melting point component are the more mobile and migrate across the interface into the higher melting point side more quickly than they are replaced by the less mobile component. Hence, vacancies are left on the side suffering a net loss of atoms. Annihilation of these defects at suitable sinks leads to a decrease in the number of lattice planes between the markers and the end of the specimen. Similarly, extra lattice planes are created on the side showing a net gain of atoms. Hence the markers shift relative to the ends. In Fig. 3.1 the shift would be from left to right.

Further discussion of the Kirkendall effect in terms of Fick's law is given in section 3.10.

Diffusion coefficients generally conform closely to an Arrhenius-type relation

$$D = D_0 e^{-E_A/kT} \qquad (3.16)$$

D_0 the frequency factor and E_A the empirical activation energy are determined by plotting log D against $1/T$ and measuring the gradient and the intercept on the log D axis (cf. 2.8).

The slight deviations from Arrhenius behaviour which are occasionally observed can usually be attributed to the intervention of grain boundary diffusion. The value of E_A for lattice diffusion is larger than for grain boundary migration in the same system. Thus the rate of decrease of grain boundary diffusion with temperature is very much lower than for lattice diffusion, and unless the grain size of a polycrystalline sample is very large it is possible that at low temperatures grain boundary diffusion makes a significant contribution to the observed flow despite the fact that the grain boundary volume is small. In this case the low temperature gradient tends to that characteristic of the value of E_A for grain boundary diffusion and at high temperature the gradient is that for lattice diffusion.

3.5. General Kinetic Theory of Diffusion in Cubic Metals

For all cubic lattices the diffusion coefficient is given by

$$D = \tfrac{1}{6}.b^2.v_J.f \tag{3.17}$$

v_J, the frequency with which an atom makes random jumps, is equal to the number of crystallographically identical jumps available to an atom times the frequency of any one jump, denoted by τ

$$v_J = \tau n \tag{3.18}$$

where n, the number of equivalent paths, is determined by lattice geometry and the mechanism of diffusion.

The simple kinetic theory (Chapter 1) may be used to express τ in terms of the free energy of activation for the movement of an atom G_{A_1}

$$\tau = v e^{-G_A/kT} \tag{3.19}$$

where v is the lattice vibration frequency. Combining eqns. (3.17)–(3.19) and subdividing G_A into the energy and entropy components (see eqns. (1.40) and (1.41)) gives a general expression for the diffusion coefficient in cubic crystals

$$D = \tfrac{1}{6}.b^2.v.n.f.e^{S_A/k}.e^{-U_A/kT} \tag{3.20}$$

The application of eqn. (3.20) to the various types of diffusion is considered in the following sections.

3.6. Diffusion of Interstitial Solutes in b.c.c. Metals

This is the easiest example to treat theoretically because (a) size factor considerations leave little doubt that solute diffusion occurs by atoms jumping from one interstitial site to an adjacent one, and (b) the maximum solubility of interstitial solutes is generally very small (e.g. C in α–Fe) and so only a small fraction of the available sites are actually occupied. Size factor considerations ensure that the occupied sites are well dispersed and thus the possibility of correlated jumps may be neglected; hence $f = 1$.

The sites occupied by interstitials in b.c.c. metals are located at the centres of the cube edges and faces as shown in Fig 3.2. They are crystallographic equivalent sites in the sense that they all have identical environments. The distance between adjacent sites is $a/2$, a being the lattice parameter, and any one atom can jump in any one of four directions as shown in Fig. 3.2. This number n is, of course, independent of temperature. Thus, for this case eqn. (3.20) reduces to

$$D = \tfrac{1}{6}.a^2.ve^{S_A/k}.e^{-U_A/kT} \qquad (3.21)$$

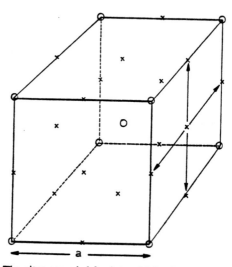

FIG. 3.2. The sites occupied by interstitial solute atoms in a b.c.c. crystal. Solute positions are marked x and solvent atoms by o. The arrows indicate the possible jump directions for a solute atom.

where S_A and U_A characterize the movement of the interstitial solute from one equilibrium site to another. The variation of a and v with temperature is small compared with the variation of the exponential term and so to a good approximation

$$\frac{d(\ln D)}{d(1/T)} = -\frac{U_A}{k}$$

But $-\mathrm{d}(\ln D)/\mathrm{d}(1/T).\mathbf{k}$ is also equal to the empirical activation energy E_A, eqn. (3.16), and so

$$E_A = U_A \tag{3.22}$$

and D_0 is given by

$$D_0 = \tfrac{1}{6}.a^2.v.\mathrm{e}^{S_A/\mathbf{k}} \tag{3.23}$$

The electronic structure of the transition metals is so imperfectly understood and so little is known about the form of interstitial solutes in these metals that as yet no attempt has been made to calculate U_A and S_A from fundamental principles. However, an empirical approach due to Zener and Wert† has proved very successful in accounting for the experimental values of S_A and D_0. G_A, the free energy of activation, is the isothermal work done in moving an atom from an equilibrium position to the top of the adjacent potential energy barrier. A local dilation of the lattice has to be induced (by thermal fluctuations) to create a void in the solvent lattice sufficient to permit the passage of the interstitial solute atoms. The work done is largely associated with the elastic strain energy of this distortion; its magnitude depends upon the relative size of the solute and solvent atoms and the elastic constants of the solvent.

The standard thermodynamic formula relating entropy and free energy is

$$S_A = -\frac{\partial G_A}{\partial T} \tag{3.24}$$

If it is assumed that all the work, G_A, is associated with the elastic strain energy, then from elasticity theory

$$G_A \propto \mu \tag{3.25}$$

where μ is the shear modulus of the lattice. To avoid having to estimate the unknown proportionality factors in (3.25) it is convenient to take the ratio G_A/G_A^0, where G_A^0 is the free energy of activation at 0°K.

---

† See paper by C. Zener in *Imperfections in Nearly Perfect Crystals* (Ed. W. Shockley), Wiley, 1952.

Hence

$$\frac{G_A}{G_A^0} \simeq \frac{\mu}{\mu_0}; \qquad (3.26)$$

μ_0 is the shear modulus at 0°K. Equation (3.26) is exactly satisfied if the proportionality factor in eqn. (3.25) is independent of temperature.

Combining eqns. (3.24) and (3.26) gives for the entropy of activation

$$S_A \simeq -G_A^0 \frac{\partial}{\partial T}\left(\frac{\mu}{\mu_0}\right) \qquad (3.27)$$

At 0°K, G_A^0 is identical with the enthalpy of activation, which is approximately equal to the internal energy of activation U_A^0. Neglecting any temperature variation of U_A (so that $U_A = U_A^0$ at all temperatures, cf. section 1.7) and using (3.22) allows (3.27) to be rewritten as

$$S_A \simeq -E_A \cdot \frac{\partial}{\partial T}\left(\frac{\mu}{\mu_0}\right) \qquad (3.28)$$

The elasticity modulii decrease as the temperature increases so that $(\partial/\partial T)(\mu/\mu_0)$ is negative, the entropy is always positive and the entropy factor $e^{s/k}$ is greater than unity.

An alternative form of (3.28) is often used, obtained by multiplying numerator and denominator by T_F, the melting temperature in °K.

Thus
$$S_A \simeq \frac{-E_A}{T_F} \cdot \frac{\partial(\mu/\mu_0)}{\partial(T/T_F)}$$
$$\simeq -\frac{E_A}{T_F} \cdot \beta \qquad (3.29)$$

where β is the dimensionless quantity

$$\frac{\partial(\mu/\mu_0)}{\partial(T/T_F)}$$

The value of D_0 for C in α-Fe calculated by combining eqns. (3.23) and (3.29) is 0·026 cm²/sec. compared with the experimental value of 0·020 cm²/sec. Despite the simplified nature of the model and the many approximations made this treatment of interstitial diffusion is regarded as one of the most satisfactory quantitative analyses of the kinetics of a process yet carried out. Such close agreement between theory and experiment, which has also been found for the diffusion of other interstitial is rarely, if ever, obtained in similar calculations of other rate processes.

3.7. Self Diffusion in Cubic Metals

The nature of the basic atomic process responsible for self-diffusion is not obvious. Four models have emerged as worthy of detailed consideration. These are illustrated schematically in Fig. 3.3.

(a) *Place exchange or ring diffusion*. In this model the diffusing atom moves by interchanging positions with another atom on a normal lattice site by the correlated rotation of two or more atoms. A two atom ring and four atom ring are illustrated in Fig. 3.3. The flux of atoms in one direction is balanced by that in the opposite direction and thus if this mechanism is operative in a two component alloy the migration of the components would be of equal velocity and in opposite directions.

(b) *Interstitial diffusion*. Diffusion results from the presence of interstitial point defects. The diffusing atom moves either through interstitial space or by displacing a neighbouring atom from a normal site to an interstitial position. In an alloy the two components have different activation energies for the formation of these defects and hence there are different numbers of each component in defect sites and also the rate of movement of the two types of interstitial are unequal. Consequently, the diffusivities of the components differ.

(c) *Vacancy diffusion*. The diffusing atom exchanges position with a vacant lattice site. The rate of jumping into a vacancy could be different for the components of an alloy permitting unequal diffusivities.

(d) *Relaxation diffusion*. The atom moves by re-arrangement within a locally disordered region produced by relaxation around one or more point defects.

The Kirkendall effect is particularly important in relation to the theories of diffusion mechanisms. Its existence shows that the rate of flow of atoms in one direction across a lattice plane is not balanced by the flow in the opposite direction. As noted above all interchange mechanisms produce equal and opposite movements

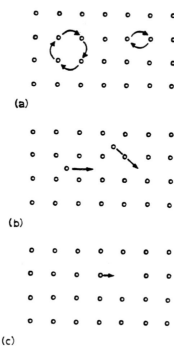

Fig. 3.3. Schematic illustration of possible mechanisms of self-diffusion: (a) a two atom ring and a four atom ring, (b) two mechanisms involving interstitial defects, (c) vacancy diffusion.

of atoms and thus it is possible to eliminate these mechanisms for all those alloys for which the effect has been observed. These include Cu–Ni, Cu–Zn, Cu–Au, Ag–Pd, Ni–Co, Ni–Au, Fe–Ni which are all f.c.c. and β-brass, U–Zr, U–Mo, U–Ti, Mo–Ti, Ti–Cr, Ti–Mo which are b.c.c.

DIFFUSION IN METALS

Kirkendall movements are only observed in alloys and thus they do not provide any direct evidence of the nature of the diffusion mechanism in pure metals. However, if diffusion is defect controlled over a wide range of compositions in a binary system it is plausible to suppose that atomic movement in the two pure metals occurs by the same mechanism, implying that diffusion in all b.c.c. and f.c.c. metals is by either vacancy or interstitial movement. To distinguish between these two it is necessary to compare measured activation energies and frequency factors and those calculated for the two possible mechanisms.

The Magnitude of the Activation energy E_A.

For self-diffusion by the vacancy mechanism an atom can leave its site only if there is a vacancy in its first-coordination shell. If all its first-nearest neighbours are vacancies the number of equivalent jump paths available (n in eqn. (3.20)) is the co-ordination number of the lattice, denoted by z. On the average, the fraction of these z sites vacant is equal to the fractional atomic concentration of vacancies in equilibrium in the crystal. Statistical thermodynamics gives the fractional concentration of point defects at a temperature T as

$$n_D/N = e^{S_F/k} \cdot e^{-U_F/kT} \quad (3.30)$$

where n_D is the number of defects in a crystal containing N atoms and S_F and U_F are the entropy and energy for the formation of one defect.

On the assumption that there is no relaxation around the vacancy the jump distance b is equal to the distance between nearest neighbours. In f.c.c. lattices $z = 12$ and $b = a/\sqrt{2}$; in b.c.c. $z = 8$, $b = (\sqrt{3}/2)a$; and in simple cubic, $z = 6$, $b = a$.

Thus, in all cubic crystals eqn. (3.20) becomes

$$D_{\text{VAC}} = a^2 \cdot v \cdot f \cdot e^{(S_F + S_M)/k} \cdot e^{-(U_F + U_M)/kT} \quad (3.31)$$

where the suffix M has been introduced to indicate the energy and entropy of activation for movement of the vacancy through the lattice.

The correlation factor for vacancy diffusion has been calculated by Compaan and Haven† to be 0·80 for f.c.c. and 0·78 for b.c.c. and is independent of temperature. The variation of a and ν with T is small compared with the exponential. Hence E_A is given by

$$E_A = -\left(\frac{d \ln D}{d(1/T)}\right)k = (U_F + U_M) \qquad (3.32)$$

Analogous considerations apply to diffusion via interstitial defects. In this case the number of first neighbours to a defect is not the lattice coordination number z but some other geometric factor determined by the lattice and the location of the interstitial position. Thus, the only modification to eqn. (3.31) is the introduction of a geometric constant. Equation (3.32) still applies. Thus, in general, the activation energy for self-diffusion is the sum of the activation energies for the formation and the movement of the particular crystal defect involved in the process.

U_F and U_M for interstitials and vacancies in copper have been calculated.‡ For vacancies U_F is in the range 0·9 to 1·2 eV and U_M 0·6 to 1·0 eV, so that E_A should be 1·5 to 2·2 eV. For interstitials U_F is between 5 and 9 eV and U_M 0·2 and 0·5 eV giving E_A 5·2 to 9·5 eV. For a two-atom ring E_A is calculated to be ~ 10 eV and about 4 eV for a four-atom ring. Evidently vacancy diffusion is the more likely on energetic grounds. The experimental value of E_A is 2·1 eV in agreement with that calculated for vacancy diffusion. No comparable calculations have been carried out for any other metal but values of U_F and U_M for vacancies have been measured in some other f.c.c. metals and generally good agreement is found between $(U_F + U_M)$ and the activation energy for self diffusion. No reliable measurements have yet been made on any b.c.c. metal, and so the position is less clear than in f.c.c. crystals. However, it is known that interstitial and ring diffusion are more likely in b.c.c. than f.c.c. crystals because the

---

† K. Compaan and Y. Haven, *Trans. Faraday Soc.* **52**, 786 (1956).
‡ H. B. Huntington, *Phys. Rev.* **61**, 325 (1942); ibid. **91**, 1092 (1953); H. Brooks, *Impurities and Imperfections*, Am. Soc. Metals, 1955; G. Fumi, *Phil. Mag.* **46**, 1007 (1955).

ion–ion repulsion which makes a substantial contribution to the activation energies for these two mechanisms in copper, which is a full metal, is not so important in the more open b.c.c. structures and so the activation energies for interstitial and ring diffusion are likely to be much closer to that for vacancy diffusion than in f.c.c. crystals.

The frequency factor

Values of D_0 for possible diffusion mechanisms have also been calculated.† In the case of f.c.c. metals, the values predicted for vacancy diffusion are in good agreement with observed values whereas those estimated for interstitial and ring diffusion are much lower. Again the position is less satisfactory for b.c.c. metals because of the greater uncertainty in the calculations and also because in some of these metals (Cr, β–Ti, β–Zr) the experimental values of D_0 (and E_A) are anomously low. It has been suggested that ring diffusion may be operative in these metals. However, the suggestion has not found much favour because Kirkendall movements have been observed at very low solute concentration in alloys based on these metals. Consequently the generally held view is that diffusion takes place by the movement of vacancies in all metals and alloys.

3.8. Impurity Diffusion

After self-diffusion the next simplest case to consider is the migration of substitutional solute atoms in a very dilute homogeneous solid solution; this is widely referred to as impurity diffusion. The tracer diffusion coefficient of the solute may be measured by studying the penetration of a small amount of radioactive isotope of the solute into an otherwise pure specimen of the solvent metal. Although a concentration gradient is established its magnitude is so small that the complications associated with Kirkendall shifts are avoided. Furthermore, by restricting the solute concentration to these levels it is justifiable to assume that

---

† C. Zener, *J. Appl. Phys.* **22**, 372 (1952); H. B. Huntington, *et al.*, *Phys. Rev.* **99**, 1085 (1955); G. M. Pound *et. al.*, *Phil. Mag.* **6**, 473 (1961); and refs. at end of chapter.

two or more solute atoms will never be nearest neighbours. Thus, only two kinds of vacancy configuration exist in the solution; namely, those in which the defect is completey surrounded by solvent atoms and those which have one solute atom in the first coordination shell.

The value of the impurity diffusivity D_B is often significantly different from that of self-diffusion of the pure solvent D_S. D_B conforms with the Arrhenius relation and so it is possible to define an empirical activation energy for impurity diffusion E_B and a frequency factor D_{OB}.

D_B is given by

$$D_B = a^2 \cdot \nu_B f_B e^{(S_{FB} + S_{MB})/k} \cdot e^{-(U_{FB} + U_{MB})/kT} \quad (3.33)$$

ν_B being the vibrational frequency of the solute atom, f_B the correlation factor, U_{FB} the activation energy for the formation of a vacancy next to a solute atom and U_{MB} the energy for the atom to exchange with the vacancy and S_{FB} and S_{MB} the corresponding entropy terms.

The activation energy for solute diffusion E_B is

$$E_B = -k\left(\frac{d(\ln D_B)}{d(1/T)}\right) = (U_{FB} + U_{MB}) - k\left(\frac{d(\ln f_B)}{d(1/T)}\right) \quad (3.34)$$

The last term is not zero because it has been shown that the correlation factor for solute diffusion varies markedly with temperature.[†] In other words, the observed activation energy is not simply the sum of the basic activation energies as is often supposed by analogy with self-diffusion.

The activation energy for solvent self-diffusion, E_S, is given by eqn. (3.32) and so $\Delta E = E_B - E_S$ is given by

$$\Delta E = (U_{FB} - U_{FS})(U_{MB} - U_{MS}) - k\left(\frac{d(\ln f_B)}{d(1/T)}\right)$$

$$= \Delta U_F + \Delta U_M - k\left(\frac{d(\ln f_B)}{d(1/T)}\right) \quad (3.35)$$

ΔE is the dominant factor in determining the difference between D_B and D_S.

---

† A. D. Leclaire and A. B. Lidiard, *Phil. Mag.* 1 518 (1956); *ibid.* 7, 141 (1962).

Considering first differences in the activation energy for the formation of a vacancy, when ΔU_F is negative it is easier to form vacancies next to solute atoms and the vacancies and solute atoms attract each other, ΔU_F being the binding energy. This association increases the effective vacancy concentration near solute atoms so that the rate of jumping of the atoms is much increased relative to that when the defects are randomly dispersed. Vacancy–solute association was first suggested by Johnson† to account for the observations that the rate of diffusion of solutes in the primary solid solutions of copper is much faster than that of self diffusion. It is thus often referred to as the Johnson effect. When ΔU_F is positive vacancies and solutes tend to keep apart and solute atoms " see " less than the average concentration of vacancies.

The factors expected to be important in determining the size and magnitude of ΔU_F are the relative sizes of the solute and solvent atoms and their relative valencies. For example, the compressive strain field around a solute atom that is larger than the solvent atom it has replaced is partly compensated by the tensile field around a vacancy when the two associate. If the difference between the valencies of the solute and solvent is Z_B then each solute ion carries a charge, relative to a solvent ion, of $Z_B e$, e being the charge on one electron. A vacancy is effectively an ion carrying zero charge and so its charge relative to the solvent lattice is $-Z_S e$, Z_S being the solvent valency. There is an attraction or repulsion between the vacancy and the solute depending upon the sign of Z_B.

Turning now to the movement of the vacancy, an atom changing places with a vacancy has to squeeze through a group of other atoms and it is to be expected that the potential barrier associated with such a jump is different for different moving atoms. Thus, even if the defects are completely randomly dispersed the tracer diffusion coefficients of the two species are different. Factors expected to be important in determining the jump energy are atom sizes and compressibility and valency effects.

---

† R. P. Johnson, *Phys. Rev.* **56**, 814 (1939).

LeClaire[†] has successfully accounted for the diffusivities of solutes in the noble metals by calculating ΔU_F and ΔU_M in terms of an electrostatic interaction model together with the temperature dependence of the correlation factor.

3.9. The Effect of a Non-Equilibrium Concentration of Vacancies

On the vacancy model the diffusion coefficient is proportional to the concentration of vacancies. Any process which produces a vacancy concentration in excess of that in normal thermodynamic equilibrium leads to an increased value of D. A supersaturation of vacancies can be introduced into a metal by quenching from a higher temperature, by cold working or by irradiation. It is to be expected that the rates of processes which are governed by diffusion will be increased when the specimen is subjected to one of these treatments. The effect of quenching is particularly relevant to phase transformation kinetics because most transformations or precipitation phenomenon are studied by first homogenizing the specimen at a high temperature followed by quenching to a temperature below that which the particular transformation would occur during equilibrium cooling. Suppose that a metal is quenched from a temperature T_Q to T_D, then the rate of diffusion at T_D is increased by a factor

$$\frac{e^{-U_F/kT_Q}}{e^{-U_F/kT_D}}$$

over and above the rate which would be measured in a normal diffusion experiment conducted at T_D, or that obtained by extrapolation of data measured at high temperatures. The value of this enhancement factor depends upon the temperature and the value of U_F. As an example, suppose aluminium is quenched from about 630° to 130°C; U_F is 0·76 eV. If all the point defects in equilibrium at 630°C are retained by the quench the enhancement factor is $\sim 10^4$. These are typical temperatures used in studying

---

[†] A. D. LeClaire, *Phil. Mag.* 7, 141 (1962).

the age hardening of aluminium alloys and so it is to be expected that the rate of ageing will be very much faster than that predicted from diffusion data relating to the transformation temperature T_D. In practice the presence of the solute modifies the value of U_F and also it is practically impossible to prevent loss of vacancies during the quench either by annihilation or by agglomeration into larger defects and the real enhancement factor cannot be estimated precisely.

The effect produced by quenching is only transient,[†] because the supersaturation decreases rapidly by the annealing out of vacancies at sinks. The following example illustrates the point. The number of jumps made per second by a vacancy is equal to the product of the rate at which it can make one jump and the number of jump paths available to it. The latter is simply the coordination number of the lattice. The jump rate is given by the usual kinetic formula $v e^{S_M/k} . e^{-U_M/kT}$ Hence if a vacancy makes n jumps in its lifetime t

$$t = n(vz)^{-1} e^{-S_M/k} . e^{U_M/kT} \qquad (3.36)$$

Taking v as 10^{13}/sec, $z = 12$, the entropy term to be 10, and n as 10^{10}, the maximum value found in studies of the annealing out of vacancies, and U_M as 0·5 eV, the value for aluminium, gives a maximum lifetime at 130°C of 30 sec.

3.10. Chemical Diffusion and the Darken Equations

Tracer diffusion in concentrated homogeneous solutions is a problem that has not yet received a satisfactory explanation. The mutual interaction of solute atoms and the various types of atomic configurations around a vacancy, make analysis in terms of atom movements very complex. Furthermore, the " association " concept cannot have the simple meaning in concentrated alloys that it has in dilute alloys. In chemical diffusion the complications are still further increased by (a) the superimposition of a concentration gradient which means that in non-ideal solutions the

---

† W. M. Lomer, *Vacancies and other Point Defects in Metals and Alloys*, Inst. of Metals, 1957.

random movement of atoms is weighted by the variation of the thermodynamic properties with composition, and (b) by the existence of the Kirkendall effect. It is not yet possible to provide a theory of chemical diffusion in terms of the detailed mechanism involved. All that is possible is to examine the general thermodynamics of the problem. The analysis that follows was first given by Darken.†

The fundamental error in Fick's laws is the assumption that concentration differences are the driving force for diffusion. The criterion for equilibrium in multi-component systems is that the partial molar free energy (p.m.f.e.) of all the components is everywhere uniform. Diffusion of one component arises when the p.m.f.e. of that component is not equal throughout the system. A simple illustration of this was given by Darken.‡ A welded pair consisting of two steels having the same carbon content but differing in silicon content, was annealed at 1050°C for 13 days, and it was found that carbon had diffused from the high silicon to the low silicon side of the couple despite the fact that the carbon concentration was initially uniform. This is consistent with the fact that silicon raises the activity and the p.m.f.e. of carbon in austenite. Silicon being substitutionally dissolved diffuses much more slowly than the interstitial carbon and so does not diffuse significantly in the course of the experiment.

It is necessary to reframe the diffusion laws in terms of p.m.f.e. and in such a way as to allow for different diffusivities of the components thereby accounting for a Kirkendall effect. However, as pointed out in section 3.2, there are considerable advantages to be gained by retaining the form of Fick's laws by defining a chemical diffusion coefficient $\tilde{D}$. The object of the Darken analysis is to relate $\tilde{D}$ at a given composition to the tracer diffusion coefficients of the components in an alloy of the same composition.

In a mechanical system the force acting on a body is the negative of the potential energy gradient. By analogy it is

---

† L. S. Darken, *Trans. A.I.M.E.* **175**, 184 (1948).
‡ *ibid.* **180**, 430 (1949).

tempting to write that the virtual thermodynamic force acting on an atom of species A in a solution is the negative of the gradient of its p.m.f.e. $-(\partial \bar{G}_A/\partial x)$ and to place the rate of flow of one atom, j_A proportional to this force, i.e.

$$j_A = - M_A(\partial \bar{G}_A/\partial x)$$

where M_A, the drift velocity under unit potential gradient, is termed the mobility and $\bar{G}_A$ is the p.m.f.e. per atom of A. This simple view ignores the possibility that one component may also experience a driving force by virtue of the potential gradient of all the other components. Thus, the most general equations for the rates of flow of the components A and B in a solution are

$$\left.\begin{array}{l} j_A = - M_{AA}(\partial \bar{G}_A/\partial x) - M_{AB}(\partial \bar{G}_B/\partial x) - M_{AV}(\partial \bar{G}_V/\partial x) \\ j_B = - M_{BA}(\partial \bar{G}_A/\partial x) - M_{BB}(\partial \bar{G}_B/\partial x) - M_{BV}(\partial \bar{G}_V/\partial x) \\ j_V = - M_{VA}(\partial \bar{G}_A/\partial x) - M_{VB}(\partial \bar{G}_B/\partial x) - M_{VV}(\partial \bar{G}_V/\partial x) \end{array}\right\} \quad (3.37)$$

where M_{AA}, M_{AB}, etc., are constants called *mobilities*.

In writing these equations it is necessary to include the vacancies as a third component since the Kirkendall effect demonstrates that a vacancy flux exists. If now it is assumed that the vacancies are everywhere in equilibrium, $\bar{G}_V$ is constant and j_V is zero. If it is further assumed that the "cross" terms are negligible, i.e. that there is no coupling between the flux of the components then eqn. (3.37) reduces to

$$\left.\begin{array}{l} j_A = - M_{AA}(\partial \bar{G}_A/\partial x) \\ j_B = - M_{BB}(\partial \bar{G}_B/\partial x) \end{array}\right\} \quad (3.38)$$

which are identical with the intuitive assumption.

Equation (3.38) is the basis of Darken's treatment of chemical diffusion. The activity of A, a_A, is defined thermodynamically by

$$\bar{G}_A = \bar{G}_A^0 + kT \ln a_A \quad (3.39)$$

$\bar{G}_A^0$ is the p.m.f.e. per atom in some arbitrary standard state. Further, a_A is related to the atom fraction of A by

$$a_A = \gamma_A N_A \quad (3.40)$$

where γ_A is the activity coefficient.

Differentiating eqn. (3.39) with respect to x and using (3.40) gives

$$j_A = -M_{AA}kT\frac{\partial}{\partial x}(\ln N_A + \ln \gamma_A) \quad (3.41)$$

Now j_A is the rate of flow of one atom; the total flux across a plane of unit area is j_A times the number of A atoms on that plane, viz. c_A where c_A is the number of A atoms per unit volume.

Comparison with Fick's first law then gives

$$D_A^c = M_{AA}kT\frac{\partial}{\partial(\ln c_A)}(\ln N_A + \ln \gamma_A) \quad (3.42)$$

Equation (3.42) defines the *intrinsic chemical diffusion coefficient of A*, D_A^c, at the composition c_A. The superfix c is attached to emphasize that this coefficient characterizes diffusion in a chemical gradient and that it is not the same as the tracer diffusion coefficient of A, in a homogeneous alloy of composition c_A, denoted by D_A.

Now

$$N_A = \frac{c_A}{c_A + c_B}$$

since the vacancy concentration is always small ($< 10^{-3}$). On the assumption of constant molar volume $c_A + c_B$ is constant and thus $d(\ln c_A) = d(\ln N_A)$.

Equation (3.42) reduces to

$$D_A^c = M_{AA}kT\left(1 + \frac{\partial \ln \gamma_A}{\partial \ln N_A}\right) \quad (3.43)$$

Similarly for B

$$D_B^c = M_{BB}kT\left(1 + \frac{\partial \ln \gamma_B}{\partial \ln N_B}\right) \quad (3.44)$$

Application of the Gibbs–Duhen equation shows that

$$\frac{\partial \ln \gamma_A}{\partial \ln N_A} = \frac{\partial \ln \gamma_B}{\partial \ln N_B} \quad (3.45)$$

from which it follows that the ratio of the two coefficients is

$$\frac{D_A^c}{D_B^c} = \frac{M_{AA}}{M_{BB}} \quad (3.46)$$

In an ideal solution $\gamma_A = \gamma_B = 1$ for all compositions and the bracketed term in eqn. (3.43) and (3.44) vanishes giving

$$D_A^c = M_{AA}kT \quad (3.47a)$$
$$D_B^c = M_{BB}kT \quad (3.47b)$$

In view of eqn. (3.47) it is apparent that the full expression for the intrinsic chemical diffusion coefficient, eqn. (3.44), is the product of two terms; one associated with the purely random jumping which would prevail if the solution is ideal and one which expresses the extent to which these random jumps are weighted due to the departure of the solution from ideality.

The next step is to relate D_A^c and D_B^c with $\tilde{D}$, derived from a chemical diffusion experiment using the Matano method. Provided that diffusion is strictly unidirectional and that the lattice remains sound with no porosity developing, the flow of atoms in such a couple can be divided into two parts:

(a) relative to the lattice given by

$$j_A^D = -D_A^c \frac{\partial c_A}{\partial x} \quad (3.48)$$

$$j_B^D = -D_B^c \frac{\partial c_B}{\partial x} \quad (3.49)$$

(b) with the lattice relative to the ends which are regarded as fixed, as revealed by the Kirkendall effect and given by

$$j_A^L = vc_A \quad (3.50)$$
$$j_B^L = vc_B$$

where $v(xt)$ is the velocity of marker movement.

The total flux of A across a plane of unit area is then

$$J_A = j_A^L + j_A^D = -D_A^c \frac{\partial c_A}{\partial x} + c_A v \quad (3.51)$$

and similarly for component B.

Following the same procedure as in the derivation of Fick's second law (section 3.2) gives

$$\frac{\partial c_A}{\partial t} = -\frac{\partial}{\partial x}(J_A)$$

$$= \frac{\partial}{\partial x}\left(D_A^c \frac{\partial c_A}{\partial x} - c_A v\right) \quad (3.52)$$

Adding eqn. (3.52) to the equivalent one for component B, noting that $(\partial c/\partial t) = \partial/\partial t\,(c_A + c_B)$ is zero if the total number of atoms in unit volume remains constant gives

$$\frac{\partial}{\partial x}\left[D_A^c \frac{\partial c_A}{\partial x} + D_B^c \frac{\partial c_B}{\partial x} - v(c_A + c_B)\right] = 0 \quad (3.53)$$

The ends of the specimen are not affected by the diffusion and thus integration of eqn. (3.53) can be effected using the boundary condition $(\partial c_A/\partial x) = 0$; $\partial c_B/\partial x = 0$; and $v = 0$ for $x = \pm\infty$.
Hence

$$D_A^c \frac{\partial c_A}{\partial x} + D_B^c \frac{\partial c_B}{\partial x} - v(c_A + c_B) = 0 \quad (3.54)$$

Converting the concentrations from atoms per unit volume into atom fractions, N_A and N_B, gives

$$v = \frac{\partial N_A}{\partial x}(D_A^c - D_B^c) \quad (3.55)$$

This equation gives the velocity of the marker movement at any concentration N_A.

Substitution of (3.55) into (3.52) gives

$$\frac{\partial c_A}{\partial t} = \frac{\partial}{\partial x}\left[(N_A D_B^c + N_B D_A^c)\frac{\partial c_A}{\partial x}\right] \quad (3.56)$$

which is the same form as Fick's equation used to define the chemical diffusion coefficient $\tilde{D}$ relative to the Matano interface.
Hence $\tilde{D}$ is related to D_A^c and D_B^c by

$$\tilde{D} = (N_A D_B^c + N_B D_A^c) \quad (3.57)$$

To determine the intrinsic chemical diffusivities it is necessary to measure both $\tilde{D}$ and v. By measuring the marker shift x_M

DIFFUSION IN METALS

relative to the Matano interface for various annealing times it is found that x_M is usually proportional to $t^{1/2}$ as expected for diffusion phenomenon.
Hence

$$x_M = \text{const } t^{1/2} \qquad (3.58)$$

and
$$v = \dot{x}_M = x_M/2t \qquad (3.59)$$

One measurement of x_M gives v at the concentration corresponding to that at x_M. Thus for one set of markers placed at the initial interface D_A^c and D_B^c may be evaluated at one value of N_A only.

Since it is not possible to measure mobilities directly it is necessary to eliminate M_{AA} and M_{BB} from eqns. (3.43) and (3.44). The tracer diffusion coefficient of A, D_A, at the composition of interest N_A may be measured by studying the penetration of a very small quantity of an isotope of A into a chemically homogeneous alloy of that composition. Since N_A is constant the second term in eqn. (3.43) vanishes and so

$$D_A = M_{AA}kT \qquad (3.60)$$

and combining eqn. (3.60) with (3.43) gives for D_A^c at the composition N_A

$$D_A^c = D_A \left(1 + \frac{\partial \ln \gamma_A}{\partial \ln N_A}\right) \qquad (3.61a)$$

Similarly,

$$D_B^c = D_B \left(1 + \frac{\partial \ln \gamma_B}{\partial \ln N_B}\right) \qquad (3.61b)$$

Combining eqns. (3.61a) (3.61b) and (3.57) and using the Gibbs–Duhem equation gives

$$\tilde{D} = (N_A D_B + N_B D_A) \left(1 + \frac{\partial \ln \gamma_A}{\partial \ln N_A}\right) \qquad (3.62)$$

By using eqn. (3.55) for the rate of marker movement v together with eqns. (3.61a) and (3.61b) gives

$$v = (D_A - D_B) \left[1 + \frac{\partial \ln \gamma_A}{\partial \ln N_A}\right] \left(\frac{\partial N_A}{\partial x}\right) \qquad (3.63)$$

The importance of eqns. (3.62) and (3.63) lies in the fact that they relate the measurable and practically important quantities

v and $\tilde{D}$ with the fundamentally significant rate at which atoms move about on the lattice at a prescribed composition, which is related to D_A and D_B. For those few systems for which sufficient thermodynamic data is available to calculate the thermodynamic term over reasonable composition ranges, agreement with the Darken equations is found. However, it must be remembered that these equations are based upon the assumptions that the lattice remains sound, that no lateral movement occurs, that the lattice parameter is independent of composition, that the vacancies are everywhere in equilibrium and that the cross terms in the diffusion equations are negligible. The first three are never completely satisfied. The existence of porosity in diffusion zones suggests that a supersaturation of vacancies is produced. Finally, Darken and others have remarked that the neglect of the cross terms has no physical justification at all. Indeed, in other branches of physics, where a similar set of linear equations is used to describe a process, it is possible to demonstrate experimentally that the cross terms are important.

3.11. Up-hill Diffusion

The experiment carried out by Darken and described at the start of the previous section is an example of diffusion against the concentration gradient, i.e. from low to high concentration. This rather special phenomenon is given the name *negative* or *up-hill* diffusion. Negative diffusion occurs when the diffusion coefficient is negative. Using Darken's theory for the intrinsic chemical diffusion coefficient it follows that this occurs when

$$D_A^c = M_{AA}kT\left(1 + \frac{\partial \ln \gamma_A}{\partial \ln N_A}\right) < 0 \qquad (3.64)$$

It is a matter of manipulation of the thermodynamic formula to convert this expression into

$$M_{AA}N_A N_B kT \left(\frac{\partial^2 G}{\partial N_A}\right) < 0 \qquad (3.65)$$

G is the mean free energy of solution per atom. Equation (3.85) was first given by Becker.†

Many binary alloy systems that are subject to phase separation at low temperature have free energy composition curves having a region of negative curvature—see, for example, Fig. 5.9. Thus in these regions negative diffusion is possible. This is physically equivalent to saying that the alloy exhibits spontaneous segregation into regions rich in A and regions rich in B. Consequently negative diffusion is particularly important in relation to the theory of the nucleation of phase separation discussed in Chapter 5.

Further Reading for Chapter 3

JOST, W., *Diffusion*, Academic Press, 1952.
BARRER, R. M., *Diffusion in and through Solids*, Cambridge Univ. Press, 1941.
CRANK, J., *Mathematics of Diffusion*.
Atom Movements, Am. Soc. Metals, 1951.
LECLAIRE, A. D., *Diffusion in Metals*, in Vols. 1 and 4 of *Progress in Metal Physics*, Pergamon Press.
Chapters 10 and 11 in *Imperfections in Nearly Perfect Crystals* (Ed. W. Shockley), Wiley, 1952.
DARKEN, L. S. and GURRY, R. W., *Physical Chemistry of Metals*, McGraw-Hill, 1953.
SEITZ, F., *Transformations in Solids*, Wiley, 1951.
Vacancies and other Point Defects in Metals and Alloys, Inst. of Metals, 1958.
BIRCHENALL, C. E., Mechanism of Diffusion in the Solid State, *Metallurgical Reviews*, Vol. 3, Inst. of Metals, 1958.
SHEWMON, P. G., *Diffusion in Solids*, McGraw-Hill, 1963.
LAZARUS, D., *Solid State Physics*, Vol. 10, 17 (Eds. F. Seitz and D. Turnbull), Pergamon, 1960.

---

† R. Becker, *Z. Metallic.* **29**, 245 (1937).

CHAPTER 4

Phase Changes in Metals and Alloys

4.1. Introduction

The phase changes of most interest are those which occur as a result of decreasing the temperature from one at which a metal or alloy exists in stable equilibrium as a single phase, say α, to another lower temperature at which the stable state consists of either (*a*) another single phase different from α, e.g. solidification of a liquid or allotropic changes in pure metals, or (*b*) two phases one of which is α of a different composition than initially and a second new phase β, e.g. precipitation from supersaturated solid solutions or (*c*) two new phases, e.g. eutectoid reactions. Immediately after the change in temperature but prior to the start of the phase change all the atoms in any structurally perfect region are in positions defined by the lattice of the parent phase. The formation of a crystal of a new phase requires that the atoms within a certain volume undergo a co-operative fluctuation which results in them taking up positions defined by the lattice of the product phase. However, not all fluctuations that produce the new lattice are stable. In fact, fluctuations below a critical minimum size are associated with an increase in free energy and are thus unstable, and any product phase produced in such a fluctuation quickly disintegrates. The reason for this is that when a new crystal is produced within the parent crystal an interface is generated and the energy associated with this interface has to be supplied by the free energy of transformation. In the case of very small crystals the interfacial energy is greater than the transformation energy,

PHASE CHANGES IN METALS AND ALLOYS 93

with the result that there is a net increase in free energy and the new crystal has only transient existence. Such an unstable crystal is called an *embryo* of the new phase. In contrast the interfacial energy of large crystals is negligible compared with the transformation energy and the net free energy change accompanying a large fluctuation is negative, with the result that a large crystal is stable.

At some intermediate size the free energy of transformation and the interfacial energy are equal. Below some critical size the fragments of the new lattice are unstable, but lattice volumes larger than this are stable and capable of continued existence. Stable volumes are called *nuclei*. Although this is a grossly oversimplified statement of the problem of nucleation, it does illustrate the general principle common to all phase changes in condensed systems that the change cannot proceed unless there exists, or are produced by internal fluctuations of structure and/or composition, stable nuclei of the product larger than some critical size. The process by which stable nuclei are produced is referred to as nucleation.

Growth of a stable nucleus takes place by the transfer of atoms from the parent to the product lattice causing the interface to advance through the parent crystal. The few atoms which are actually in the process of change (see Chapter 1) at any instant are those located in the interface. Growth occurs more readily than nucleation because the increase in size is favoured by the tendency of the system to minimize the proportion of material in the high energy interfacial region. It follows that the transformation of the material in a considerable volume around a nucleus is accomplished by growth of that nucleus rather than by fresh nucleation.

Nearly all phase changes in solids occur by the growth of a limited number of nuclei. It is usual to classify these by consideration of the mechanism of the growth process.

4.2. Diffusional Transformations

Diffusional transformations are those in which the interface advances by the thermally activated movement of single atoms across the interface. The basic atomic process is analogous to diffusion in that it involves the uncoupled jumping of individual atoms through distances comparable to the interatomic distance in either lattice. When a composition change is involved it is accomplished by the various kinds of atoms being transported to or from the interface through parent and/or product over distances large relative to the interatomic spacing. Eutectoid formation and precipitation reactions are examples of this type. Polymorphic transformations in pure metals may also be diffusional, although no composition change is involved. In this case it is unlikely that atoms move through long distances in the parent but provided the interface reaction involves the thermally activated movement of single atoms they are properly classified in this category.

It is useful to distinguish between two broad types of diffusional reactions in multicomponent systems. In the first, known as *discontinuous precipitation*, the structural and compositional changes occur in regions immediately adjacent to the advancing interface. The parent phase remains unchanged until swept over by the interface and the transition is complete in regions over which the interface has passed. After a steady state has been realized the rate of growth is constant until two regions of product impinge when the rate decreases abruptly to zero. Eutectoid formation is in this class. Polymorphic transitions in pure metals are a limiting case wherein the composition change is zero and the structural change is accomplished at the interface. In the second type, called *continuous precipitation*, atoms are transported to the growing nuclei by diffusion over relatively large distances in the parent phase. The mean composition of the parent phase changes continuously towards its equilibrium value. The structural change, if any, is localized at the interface. The rate of growth depends upon the relative rates of the interface reaction and diffusion and will only be constant if the former is very much slower than the latter. Precipitation from many supersaturated solid solutions involves continuous precipitation.

In general, diffusional phase changes occur at high temperature where atoms are sufficiently mobile. The rate is very sensitive to temperature: part of this dependence arises from the variation of the growth rate with temperature which may in itself be complicated since it involves more than one thermally activated step and because the driving force (the free energy change) varies with temperature; and part is associated with the temperature dependence of nucleation which is also complex. Thus, it is not possible to generalize about the effect of temperature on the overall reaction rate. Because these reactions require time at high temperature it is possible to completely or partly suppress them by rapid cooling.

Diffusional changes do not necessarily produce the most stable arrangement. The only thermodynamic condition for any change is that it is accompanied by a decrease in free energy. Several important transformations are known in which the most stable state is achieved via several transition stages each involving the formation of metastable phases of increasing stability. Precipitation from supersaturated aluminium alloys and the tempering of martensite are notable examples.

4.3. Martensitic, or Shear Transformations

When the decomposition of a phase by a diffusional process is suppressed by means of quenching, the parent phase is retained in metastable equilibrium to low temperatures. The driving force tending to make the phase decompose increases as the undercooling increases and in many systems it becomes sufficiently high to permit an alternative mode of transformation. In this, once the stable nucleus is formed, the interface is propagated by means of the systematic coordinated shear of large regions of the parent phase containing many atoms in such a way as to create an entirely new lattice. The distance through which any single atom moves relative to its neighbours is less than one lattice spacing. No interchange of position is possible and so the composition of the parent and product phase is identical. For this reason these phase changes are termed *diffusionless*, or *shear* transformations. The term martensitic is also used frequently because the best known and technically the most important example is the decomposition of quenched austenite at temperatures near to room temperature to form the constituent martensite. This constituent is responsible for the characteristic hardness and micro-structure of quenched steel.

The shear of a volume of crystal to produce a new lattice containing the same number of atoms must involve a change in the shape of the volume. The shape change gives rise to considerable elastic strains in product and matrix. The energy associated with this distortion has to be supplied by the free energy difference between the two phases. The reactions occur only in cases of drastic undercooling where the free energy change (the driving force) is large. In fact there is generally a temperature above which a martensitic reaction is not observed due to insufficient driving force. This temperature is designated M_S.

Martensite in steels is a metastable phase differing in structure and/or composition from that of the equilibrium phase. Many other examples of this kind are known. There are also systems in which a single equilibrium phase can be produced by a diffusionless transformation because the temperature of reaction is below the two phase region in the equilibrium diagram.

4.4. Intermediate or Mixed Transformation Mechanisms

Pure diffusional or shear transformations are limiting cases, the former occurring at high temperature where atomic mobility is appreciable but the undercooling, and hence the driving force, small and the latter at low temperature where mobility is negligible but the driving force considerable. Intermediate or mixed types are to be expected at intermediate temperatures where mobility is small, but not zero and where the driving force is inadequate to sustain a completely shear type process. At least two have been recognized experimentally:

(i) The "massive" reaction observed in some copper and some iron base alloys. These alloys undergo a diffusional change at high temperature and a pure martensitic transformation at low temperature. A third mode occurs in a range of temperatures just above the temperature range of the shear action giving a product in the form of massive irregular grains with facetted boundaries. The presence of a shape change and the chemical identity of product and parent indicate that the transformation is essentially diffusionless. The widely accepted view is that the temperature is sufficiently high to permit movement of individual atoms over two or three lattice spacings which enables relaxation of the lattice strains to proceed continuously by means of short range diffusion or thermally activated dislocation rearrangement, thereby reducing the forces opposing the shear.

(ii) The bainite reaction in steels. Bainite is a fine dispersion of ferrite and carbide produced by the decomposition of austenite at temperatures between the martensite range and that at which pearlite is produced diffusionally. The fact that phase separation occurs indicates that diffusion is an integral part of the reaction process. In addition a shape change is observed within the ferrite matrix, demonstrating that lattice shear is also involved. This type of change is to be expected in systems in which the activation energy for diffusion of one component is much less than that for the others. In the case of iron base alloys the interstitially dissolved carbon migrates with an activation energy in the order of 20 kcal/mole, compared with a value of about 50 kcal/mole for iron. At temperatures in the range 200–400°C the carbon atoms have appreciable mobility though the iron atoms are relatively immobile. Thus, transformation of the iron lattice must occur by shear but redistribution of the carbon atoms within the parent or product is possible by diffusion to give carbide and ferrite. Because the carbide–ferrite aggregate has a lower free energy than the product of a pure diffusionless change the diffusional component makes available additional free energy that enables the shear to proceed at a temperature above M_S at which the driving force would otherwise be inadequate.

CHAPTER 5

Nucleation

5.1. Introduction

No experimental technique has yet been devised that permits direct observation of the formation of a nucleus. Consequently, it is necessary to resort to indirect means of study. The general procedure is to assume a model of a nucleus, to calculate the rate at which such a nucleus should form as a function of temperature, composition and other relevant variables and then to test the validity of the model by comparison with experimental measurements of nucleation rate. The major obstacle to progress with this approach is that of measuring the rate of formation of nuclei when there is no means of observing them. It is appropriate to start the chapter by considering how the experimental problem can be overcome.

The nucleation rate I is defined as the number of nuclei of product phase formed per unit time per unit volume of parent phase. One method used to estimate I is to count the number of discrete volumes of product after they have been allowed to grow to an observable size, using either electron or optical metallography. Since the observations are made on a plane surface, and I is expressed in terms of unit volume it is necessary to determine not only the number but the size distribution of particles. Procedures are available for deriving the number of particles in unit volume from the number counted on a random plane, but they are based on simplifying assumptions and the results are subject to large errors. Measurements on a series of samples reacted for various times gives the number of particles as a function of time from which I is readily estimated. To date, the measurements of nucleation frequency by this method have been attempted only for a few systems in which the product grows to a size observable by optical microscopy. An implicit assumption is that the time required for a particle to grow from the critical nucleus size to that at which it is observed is independent of the time at which it formed, i.e. that the time dependence of the growth rate is just the same for the first formed as the last formed particles. It is a plausible, but not an unquestionable assumption for discontinuous precipitation and eutectoid changes, because in these reactions a point in the matrix is completely unchanged until swept over by a product–parent interface; the same applies to diffusionless reactions. But it is certainly not true for continuous precipitation processes in which the matrix is continuously drained of solute from the time the first nucleus is formed.

A more readily applied method is based upon the assumption that the time t_y, required for a specified fraction y of the parent phase to transform, is inversely proportional to the initial rate of nucleation. The temperature dependence of I is then obtainable by measuring t_y at a series of temperatures. A serious objection to the method is that t_y depends upon the growth velocity as well as nucleation rate and the temperature dependence of growth is usually quite different from that of nucleation. Further, growth and nucleation are usually different functions of y and thus the temperature dependence of t_y is different for different values of y. The most valid application of the method is to that class of diffusionless changes in which growth rate is not only extremely rapid but virtually independent of temperature (see Chapter 8). In such cases the rate of transformation is determined by the rate of nucleation only and so the time and/or temperature dependence of y (and so t_y) reflects that of I.

A third method occasionally used is to derive or assume a functional relationship between the overall reaction rate and the rates of nucleation and growth. Growth rates are then measured metallographically as discussed in Chapter 6, and combined with rates of reaction to give nucleation rates by difference. The reliability of the results depend upon the validity of the rate equation used, but this is rarely known with confidence because proof requires measurement of nucleation rates.

5.2. Homogeneous and Heterogeneous Nucleation

Nucleation that occurs completely at random throughout a system is said to be *homogeneous*. A condition for homogeneous nucleation is that any small volume element in the parent phase is structurally, chemically and energetically identical with every other element. This is possible only if the specimen is chemically homogeneous and free from structural imperfections. In practice solids contain many imperfections such as surfaces, grain boundaries, impurity particles and dislocations and it is doubtful if completely homogeneous nucleation can ever be obtained in practice. The nearest approach is realized in the droplet experiments discussed in section 5.4. In real crystals the energy of a group of atoms depends upon the location, being greater for groups of atoms at structural imperfections than for atoms in perfect regions. Thus, the energy required to generate a nucleus is generally less if the nucleus forms at one of these high energy locations and, consequently, nucleation tends to be associated with these sites. Nucleation at preferred sites is designated *heterogeneous nucleation*.

The theories of nucleus formation are first presented in relation to homogeneous nucleation. The extension to heterogeneous nucleation presents no new concepts and is conveniently made later.

5.3. The Classical Theory of Homogeneous Nucleation

At temperatures at which atomic mobility is appreciable local rearrangements of the various atomic species occurs continuously as a result of thermal agitation. If the phase is thermodynamically stable such regions have only transitory existence and rapidly disperse to be replaced by others at some other location. However, if the phase should be metastable these fluctuations become of considerable importance because they are a potential source of nuclei for a change to a more stable structure. Embryos produced by thermally activated fluctuations may vary in size, shape, structure, composition (if the system has two or more components) and may or may not be internally uniform, structurally and chemically. The simplest model is to assume that embryos are internally uniform and have the same structure, composition and properties as the final product phase in bulk form. This is the basis of the classical theory of nucleation originated by Volmer and Weber and Becker and Döring† for condensation from vapours and since applied to all types of phase changes.

These assumptions leave the shape and size of the embryo or nucleus as the only variable parameters. The shape is that which minimizes the energy of formation. The energy expended in the formation of a nucleus consists of (a) the energy of the interphase boundary, (b) elastic strain energy arising from any change in volume accompanying the structural change, and (c) lattice strain energy associated with the lattice distortion produced by partial or complete coherency between the two lattices. The optimum shape is closely linked with the nature of the interphase interface and the existence of crystallographic relationships between the phases. A discussion of the details is best left to a separate section. For the present it is assumed that the interfacial energy is independent of crystallographic orientation and that the strain energy is negligible. On the basis of these assumptions the requirement of minimum surface energy results in spherical embryos.

The second variable, size, is determined from the thermodynamic condition for stability. The free energy of formation of an embryo of radius r, of a phase β within a phase α is

$$\Delta G = 4/3\pi r^3 \Delta G_v + 4\pi r^2 \gamma \qquad (5.1)$$

in which ΔG_v is the difference between the free energy of α and β per unit volume of β measured on bulk samples, γ is the interfacial energy per unit area of the α–β interface expressed in appropriate units, and assumed independent of r. The second term in eqn. (5.1) is always positive. If the temperature is such that α is stable

---

† M. Volmer and A. Weber, *Z. Phy. Chem.* **119**, 277 (1925); R. Becker and W. Döring, *Ann. Phys.* **24**, 719 (1935).

relative to β the first term is also positive and ΔG is positive and increases rapidly with r. Embryos of all sizes are unstable. However, statistically there is a steady state distribution of sizes, the individual embryos of which are in the process of growth or dispersion. The distribution is characteristic of the temperature.

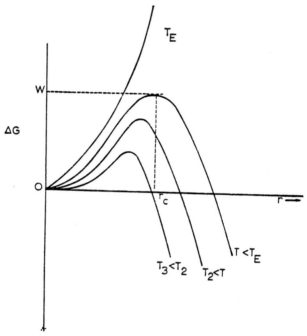

Fig. 5.1. The free energy of formation of spherical embryos as a function of the radius for a series of temperatures.

When β is stable relative to α, ΔG_v is negative. At small values of r the surface term dominates and ΔG is positive; at large r the volume free energy term dominates because this is proportional to r^3. ΔG passes through a maximum, denoted by W, at a radius r_c. This result is shown graphically for a temperature T below T_E in Fig. 5.1. The value of W and r_c depend upon ΔG_v

and thus upon the temperature. As the temperature is decreased below T_E, ΔG_v becomes increasingly negative and W and r_c decrease as shown in Fig. 5.1. At T_E, W and r_c are both infinite.

r_c is the *critical nucleus size*. Growth of embryos smaller than r_c leads to an increase in the free energy and thus there is a greater tendency for such embryos to shrink rather than grow; embryos larger than r_c are stable because growth is accompanied by a decrease in G. Embryos of radius r_c are unstable having an equal chance of shrinking or growing.

W the work, or free energy, of formation of a stable nucleus is that for which

$$\frac{\partial(\Delta G)}{\partial r_{r=r_c}} = 0 \qquad (5.2)$$

Differentiation of eqn. (5.1) with respect to r and the application of eqn. (5.2) gives

$$r_c = -\frac{2\gamma}{\Delta G_v} \qquad (5.3)$$

and

$$W = \frac{16}{3}\pi \cdot \frac{\gamma^3}{(\Delta G_v)^2} \qquad (5.4)$$

It follows from eqn. (5.4) that the free energy of formation of a stable nucleus is infinite at T_E. Physically this means that it is impossible for a phase change to occur at the equilibrium temperature. This is in accord with the experience that undercooling is always necessary for a phase change. Applying the same argument it follows that superheating to above T_E is necessary for a heating phase change. Consequently phase changes occur at different temperatures on heating and cooling. The temperature at which the phases are in thermodynamic equilibrium is somewhere between these two. This effect is known as transformation *hysteresis*.

The next step is the calculation of the size distribution, i.e. the number of embryos per unit volume as a function of their size. The problem is much simplified by the use of two assumptions. The first is the basic one of quasi-equilibrium kinetics that virtual equilibrium exists and that this is not distorted by the continuous removal of critical size nuclei from the distribution. The second is that a group of embryos of given size behave independently of all other size groups, which permits the calculation of the number of a given size without reference to any other.

Let there be N sites per unit volume at which embryos can form and the number of embryos of radius r per unit volume be n_r. The equilibrium between N sites and n_r embryos is characterized by an equilibrium constant, K, with

$$K = \frac{n_r}{N} \tag{5.5}$$

K is related to ΔG the free energy of formation by the standard equation

$$K = e^{\Delta G/kT}$$

Therefore

$$n_r = N e^{\Delta G/kT} \tag{5.6}$$

The equilibrium number of nuclei of critical size n_c per unit volume is

$$n_c = N e^{-W/kT} \tag{5.7}$$

An embryo of size r_c becomes a nucleus when it gains one or more atoms. If the jump process across the interface is governed by an activation energy U_I, the rate of interface movement is proportional to $e^{-U_I/kT}$ and the frequency with which critical embryos become stable is

$$n_s^* . p . v . e^{-U_I/kT} \tag{5.8}$$

where n_s^* is the number of atoms in the matrix at the surface of the critical embryo, v is the frequency of vibration of these atoms, and p is the probability that a vibration is in the direction of the embryo times a factor which expresses the fact that attachment of atoms may only occur preferentially at certain points on the surface. In the Volmer-Weber theory it is assumed that (a) loss of critical nuclei due to growth is balanced by formation of new ones, (b) the resulting steady-state number is also the equilibrium number n_c and (c) shrinkage of nuclei larger than r_c is negligible. I, the number of nuclei that appear per unit volume of parent per unit time under steady state conditions is then

$$I = N . n_s^* . p . v . e^{-U_I/kT} . e^{-W/kT} \qquad (5.9)$$

$$= A\, e^{-(U_I + W)/kT} \qquad (5.10)$$

where A, the frequency factor, is the product of the pre-exponential terms in eqn. (5.9).

5.4. Nucleation of Crystals from Pure Liquid Metals

The phase change that comes nearest to satisfying the assumptions made in the classical theory is the solidification of a pure metal. No composition change is involved. The strain energy associated with a nucleus of solid growing in a liquid is negligible and there is little chance of crystallographic relationships between solid and liquid. The free energy of a solid and liquid phase as a function of temperature is shown in Fig. 5.2. Since over fairly small ranges of temperature both relationships are nearly linear it is a reasonable approximation to put the difference $(G_s - G_L)$ proportional to the degree of undercooling $(T_E - T)$. Thus the volume free energy change on solidification, ΔG_v is given by

$$(\Delta G_v) = \text{const}\,(T_E - T) \qquad (5.11)$$

The standard thermodynamic expression for ΔG_v at any temperature is

$$\Delta G_v = \Delta H_v - T\Delta S_v \qquad (5.12)$$

ΔH_v and ΔS_v being the difference in enthalpy and entropy between solid and liquid per unit volume.

Assuming that ΔH_v and ΔS_v are independent of temperature, ΔS_v can be evaluated by noting that at the freezing point T_E the entropy change is equal to ΔH_v divided by T_E. Thus the volume free energy change at any temperature T is given by

$$\Delta G_v = \Delta H_v - T \cdot \frac{\Delta H_v}{T_E} = \frac{\Delta H_v}{T_E}(T_E - T) \quad (5.13)$$

Combining eqns. (5.4) (5.7) and (5.13), the number of spherical nuclei of critical size is

$$n_c = N \exp\left[-\tfrac{16}{3}\pi\gamma^3 \frac{T_E^2}{(\Delta H_v)^2(T_E - T)^2 kT}\right] \quad (5.14)$$

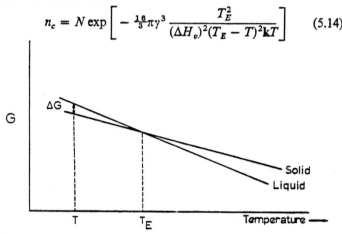

FIG. 5.2. The free energy of solid and liquid phases. T_E is the melting point and ΔG the free energy change when solidification occurs at temperature T.

Equation (5.14) can be used to estimate the undercooling required to produce spontaneous solidification during continuous cooling. For freezing to begin a reasonable number of nuclei per unit volume is necessary and the required undercooling is that value of $(T_E - T)$ which will produce in eqn. (5.14) a value of n_c of say unity. In fact, the calculation is quite insensitive to the value of n_c assumed because $(T_E - T)$ appears in the exponential. Taking N as 10^{23} gives

$$\ln 10^{23} = \tfrac{16}{3}\pi\gamma^3 \frac{T_E^2}{(\Delta H_v)^2 (T_E - T)^2 kT} \quad (5.15)$$

Theoretically, the solid–liquid interfacial energy for close-packed metals is expected to be of the order of 200 erg/cm². Substitution of this value into eqn. (5.15) using appropriate units gives a value of approximately 200°K for the undercooling required for solidification.

In practice bulk samples (> 1 cm³ in vol.) can rarely be made to undercool more than 5°K. The reason for the disparity is that actual samples always contain an abundant supply of sites at which heterogeneous nucleation can occur. In order to test eqn. (5.15) it is necessary to devise an experiment in which heterogeneous nucleation is obviated. Turnbull achieved this by dividing the liquid into a large number of droplets, 10–50μ in diameter, separated from each other by being suspended in oil or slag or by being spread on a plate.† The idea is that if the number of droplets is greater than the number of the foreign particles which act as heterogeneous nucleation sites then some of the droplets will be free of particles and be available for homogeneous nucleation. Solidification was observed either by microscopic examination or dilatometrically. The droplets were found to solidify in groups, each group at a sharply defined and reproducible temperature but each group at a different temperature. This suggests that the bulk sample contained various types of nucleating agent, each of different effectiveness. Droplets containing particles of the most effective catalyst solidify as a group at the highest temperature, those containing the second most effective catalyst solidify at a slightly lower temperature and so on. The group solidifying at the lowest temperature is that in which there are no effective foreign particles and in which homogeneous nucleation occurs.

The technique was applied to a wide selection of metals and in most cases it was found possible to achieve maximum undercooling of up to two or three hundred degrees, as expected from eqn. (5.15). Since the solid–liquid interfacial energy of most of the metals was not known exactly it was not possible to compare the results with calculated degrees of undercooling. Instead the measurements were used with eqn. (5.15) to obtain values of γ.

---

† D. Turnbull, *J. Chem. Physics*, **18**, 769 (1950); *J. App. Phys.* **21**, 1022 (1950); *ibid.* **20**, 817 (1949); *Acta. Met.* **1**, 8 (1953).

108 THE KINETICS OF PHASE TRANSFORMATIONS IN METALS

For most of the close packed metals this turned out to be in the range 150–200 erg/cm^2 which is in reasonable agreement with the value expected theoretically.

Whilst this satisfactory result supports the general basis of classical nucleation theory it can hardly be regarded as a detailed justification because the test is very insensitive. Changing the value of I from unity as assumed in deriving eqn. (5.15) to 1000 nuclei/unit vol has negligible effect on the derived value of γ.

Some further results of the droplet experiments are considered in the following section.

5.5. Heterogeneous Nucleation During Solidification

The fact that bulk samples of metals undercool less than 5°K compared with the very much larger degrees of undercooling attainable when homogeneous nucleation is realized is clear evidence of the effectiveness of foreign particles as nucleating catalysts during solidification. The theory of heterogeneous nucleation on impurity surfaces is a straightforward extension of the classical nucleation theory. The formation of a nucleus of solid on a plane surface of a particle suspended in the liquid is illustrated in Fig. 5.3. In addition to the creation of the solid–liquid interface it is necessary to consider the energies of the solid–impurity interface γ_{PS} and the liquid–impurity interface γ_{PL}. If θ is the angle of contact, balance of the horizontally resolved surface tensions requires that

$$\gamma_{PL} = \gamma_{PS} + \gamma \cos \theta \qquad (5.16)$$

Actually, the surface cannot be exactly plane because this would result in an unbalanced vertical force. However, this is readily allowed for and does not significantly change the argument.

Repetition of the derivation of the free energy of formation of a critical nucleus in the same way as in (5.3) gives

$$W = \frac{16\pi}{3} \frac{\gamma^3}{(\Delta G)_v^2} \left[\frac{(2 + \cos \theta)(1 - \cos \theta)^2}{4} \right] \qquad (5.17)$$

This expression is the same as the result for homogeneous nucleation apart from the function in brackets. When $\theta = 180°$

this function equals unity and W is the same as for homogeneous nucleation. Physically this is equivalent to saying that the solid does not "wet" the particle; the nucleus has only point contact with the surface. In this case the particle is completely ineffective as a catalyst. The other limit is when $\theta = 0$ giving $W = 0$, i.e. no barrier to nucleation. $\theta = 0$ implies perfect wetting so that the solid is spread into a thin film along the surface. This could be realized perfectly if the catalyst is a crystal of the solid metal. At intermediate values of θ, W decreases with decreasing θ and the

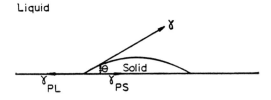

Fig. 5.3. The formation of a nucleus of a solid phase on the surface of a foreign particle.

smaller the value of W the less the supercooling required for the onset of solidification. Thus nucleating agents may be characterized by a unique value of the contact angle θ, the most effective being those with low values of θ.

Heterogeneous nucleation in liquids was studied by Turnbull using the droplet technique.† Since the surface of the droplets is the probable site for heterogeneous nucleation the effectiveness of a given substance as a catalyst can be studied by coating the droplets with the substance. The best example is the solidification of mercury droplets ($T_E = -39°C$). After coating chemically, the droplets were suspended in a liquid contained in a dilatometer. The dilatometer was placed in a thermostat and the fall in the level

---

† D. Turnbull, *J. Chem. Physics*, **20**, 411 (1952).

of the dilatometer liquid was recorded as a function of time. Some results for droplets coated with mercuric acetate are shown in Fig. 5.4. In general, it was found that solidification occurred over a narrow temperature range (2–4°C) for a given coating and that the supercooling for appreciable nucleation rate, ranging from 3° to 75°C, was primarily determined by the type of surface film.

The rate of nucleation is given by combining eqns. (5.9) (5.13) and (5.17)

$$I = n_s^* . N . p . v_L . \exp - \left[\frac{16\pi}{3} \gamma^3 \frac{T_E^2}{(\Delta H_v)^2 (T_E - T)^2 kT} f(\theta) + \frac{U_I}{kT} \right] \quad (5.18)$$

in which $f(\theta)$ is the function inside the brackets in eqn. (5.17). U_I is of the same order as the activation energy for diffusion in the liquid and since this is small $U_I{}^{kT}$ may be neglected. Thus

$$\ln I = \ln (n_s^* . N . p . v_L) - \frac{16\pi}{3} \gamma^3 \frac{T_E^2}{(\Delta H_v)^2 (T_E - T)^2 kT} f(\theta) \quad (5.19)$$

The rate of growth of crystals from the liquid is very rapid and it is reasonable to assume that the rate of solidification is determined by the rate of nucleation. For one catalysing agent I at a given temperature is related to the size of the droplet because N, the number of possible sites, is related to the surface area of the droplets. In interpreting the solidification isotherms it is thus necessary to allow for the fact that they manifest the solidification of all sizes. For details of how this was done reference should be made to the original paper. It was found that the curves in Fig. 5.4, which are for mercuric acetate coated particles, could be satisfactorily interpreted on the basis that solidification rate is proportional to the surface area of the droplets and that a single value of I exists. This is consistent with the supposition that nucleation occurs exclusively at the interface of the metal and the surface film. The derived values of $\ln I$ are shown in Fig. 5.5 as a function of

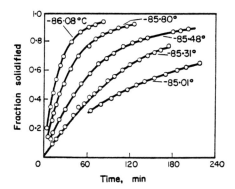

Fig. 5.4. Solidification isotherms obtained using a volume dilatometer for the solidification of droplets of mercuric acetate (Turnbull).

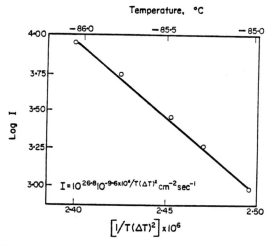

Fig. 5.5. Values of the rate of nucleation I derived from the isotherms in Fig. 5.4 plotted against $1/(T_E - T)^2 \cdot T$ (Turnbull).

$1/(T_E - T)^2 \cdot T$. According to eqn. (5.19), this plot is linear of slope

$$\frac{16\pi}{3} \gamma^3 \cdot \frac{T_E^2}{(\Delta H_v)^2} \cdot \frac{1}{k} \cdot f(\theta)$$

from which θ can be derived once γ is known. γ was evaluated from identical experiments with droplets coated with mercuric laurate. In this case it was found that solidification occurred only when the supercooling was more than 75°C, and that the rate was proportional to the droplet volume, indicating that nucleation was homogeneous, i.e. $\theta = 180°$. Combination of the two results gave $\theta = 72°$ for the contact angle between solid mercury and a mercuric acetate coating.

The factors that determine the effectiveness of a particular substance as a nucleating catalyst have not yet been positively established. It can be seen from eqn. (5.16) and Fig. 5.3 that θ decreases as the solid–substrate interfacial energy decreases. Factors that might affect this energy are (a) lattice disregistry between substrate and solid metal, (b) chemical nature of the substrate, (c) surface imperfections and (d) adsorbed films on the surface. There is considerable evidence from observations of the effect of innoculation on the grain size of cast metals that the most potent nucleating catalysts are those that possess a low index plane with a similar atomic arrangement to a low index plane in the metal lattice. This suggests that a coherent interface is formed between substrate and solid, the energy of which is very much lower than that of an incoherent solid–liquid interface.

5.6. Some Further Results of Classical Nucleation Theory

(i) *Temperature-dependence of nucleation rate*

The temperature dependence of I, given by eqn. (5.10), arises through several factors,

(a) The variation of A through the term n_s^*. This is negligible in comparison with the temperature dependence of the exponential terms.

NUCLEATION   113

(b) The term $e^{-U_I/kT}$, which since U_I is constant decreases rapidly with T, becoming zero at 0°K.

Hence I is zero at absolute zero.

(c) The term $e^{-W/kT}$ with W given by eqn. (5.4).

Surface energies vary only slightly with temperature and so γ may be regarded as constant. ΔG_v is a function of the undercooling, becoming increasingly negative as the temperature departs from the equilibrium transition temperature T_E. Assume for the purpose of illustration, that ΔG_v is directly proportional to the

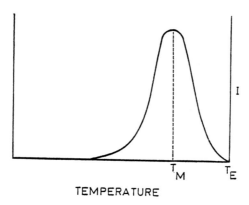

FIG. 5.6. The variation of the rate of nucleation I with temperature.

degree of undercooling, $(T_E - T)$ so that $W \propto 1/(T_E - T)^2$. Thus $e^{-W/kT}$ is zero at T_E and increases continuously as T decreases. Hence I is also zero at T_E.

Combining (b) and (c) shows that as the temperature is decreased below T_E the rate of nucleation increases from zero, passes through a maximum at some temperature T_M and then decreases to zero at 0°K as shown in Fig. 5.6. The physical interpretation of this behaviour is that at higher temperatures the initial nucleus size is so large that few are formed despite the high atomic mobility. Decreasing the temperature decreases both the size of the nucleus and the mobility. At small undercooling the decrease in nucleus size is more rapid than the decrease in mobility and this factor predominates causing nucleation rate to increase. At low temperatures the situation is reversed and despite the fact that the nucleus size is very small the atomic mobility is inadequate to permit their formation.

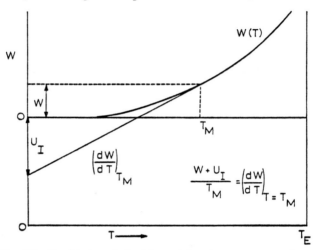

Fig. 5.7. Graphical method of determining the temperature of maximum nucleation rate from a graph of the activation energy for nucleation W against temperature.

T_M can be determined by the following graphical procedure, provided W is known as a function of T. Taking logarithms of eqn. (5.10) gives:

$$\ln I = \ln A - \frac{U_I}{kT} - \frac{W}{kT} \tag{5.20}$$

and

$$\frac{d(\ln I)}{dT} = \frac{U_I}{kT^2} + \frac{W}{kT^2} - \frac{1}{kT}\left(\frac{dW}{dT}\right) \tag{5.21}$$

T_M is the temperature at which $d(\ln I)/dT = 0$ and is thus given by

$$\left(\frac{dW}{dT}\right)_{T=T_M} = \frac{W + U_I}{T_M} \tag{5.22}$$

NUCLEATION

In Fig. 5.7 the curve gives W as a function of temperature. A length equal to U_I is laid off on the W axis in the negative direction and from this point a tangent is drawn to the curve. The point of contact is T_M.

On the assumption that the isothermal rate of a transformation is directly related to the initial nucleation rate, eqn. (5.10) predicts that the transformation rate is small both near to T_E and at low temperature and a maximum at some intermediate temperature. In other words, the *TTT* curves for phase changes should be in the form of a *C*-curve, which is in accord with experience.

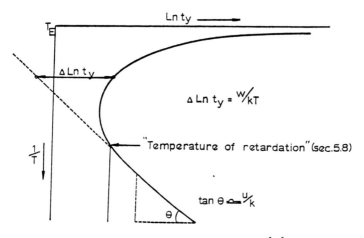

FIG. 5.8. Method of finding the values of U_I and of $\dot{W}$ as a function of temperature from an inverse temperature TTT diagram. (After Hardy and Heal.)

It is more useful to plot *TTT* diagrams in terms of the reciprocal of the absolute temperature $1/T$. On the assumption that the time t_y to a given fraction transformed y is inversely proportional to the rate of nucleation, it is possible to write

$$\ln t_y = U_I/kT + W/kT - \ln A \qquad (5.23)$$

$$k\left[\frac{d(\ln t_y)}{d(1/T)}\right] = U_I + W + \frac{1}{T}\left[\frac{dW}{d(1/T)}\right] \qquad (5.24)$$

If the change obeys C-curve kinetics the type of reciprocal rate curve is as shown in Fig. 5.8. Since at low temperature $W \simeq 0$ the low temperature part of the graph is approximately linear of slope U_I/k. It often turns out that the activation energy determined for the low temperature section of a *TTT* diagram is substantially different from that for the anticipated diffusion processes. There are several possible reasons for this (a) as emphasized previously, t_y is determined by growth as well as by nucleation and thus the low temperature slope of the reciprocal rate curve is some function of the activation energies for the two processes, (b) normal diffusion studies are generally carried out at high temperature where diffusion short circuits (paths of low activation energy) are not important, whereas at the low temperature range covered by the rate curve the short circuits account for most of the transport and (c) excess vacancies may be trapped by the quench and lead to accelerated diffusion.

In addition to obtaining U_I it also is possible to obtain experimental values of W from a reciprocal rate curve (Fig. 5.8). The low temperature linear portion is extrapolated to higher temperatures as indicated by the dotted line. This line represents the equation

$$\ln t_y = U_I/kT - \ln A \tag{5.25}$$

At any temperature the distance on the time axis between the line and the curve ($\Delta \ln t_y$) is measured. This procedure is equivalent to subtracting eqn. (5.25) from eqn. (5.23). Hence W at any temperature T is given by

$$W = kT[\Delta \ln t_y] \tag{5.26}$$

(ii) Variation of nucleation rate with time at constant temperature

The expression for the nucleation rate in eqn. (5.10) was based upon the assumption that the equilibrium distribution of embryos exists at all times. In practice this is not usually the case. Before quenching to the reaction temperature the system is equilibrated at some temperature above T_E during which time the embryo distribution characteristic of that temperature is established.

This distribution is retained by the quench and thus a time interval must elapse whilst the distribution characteristic of the new temperature is formed. If all the inherited embryos are less than the critical size at the reaction temperature an incubation period, during which no transformation occurs, is evident, being the time required for the formation of a significant number of nuclei. This is followed by a period in which the rate of nucleation increases from zero to the steady state value. Some approximate analyses of this transient nucleation have been given.† Qualitatively these ideas are consistent with experience since an incubation period is a feature of most isothermal phase changes.

The steady state rate of homogeneous nucleation should remain constant once it is established for phase changes in pure metals, for discontinuous precipitation and diffusionless transformations in alloys, because the parent phase is completely unchanged until swept over by the transformation interface. In continuous precipitation the supersaturation decreases continuously throughout the matrix, so that ΔG_v decreases continuously. In view of the exponential dependence of I on $(\Delta G_v)^2$ it is to be expected that I will decrease rapidly from its initial value and vanish early in the transformation. No quantitative treatment has been attempted but again it is of interest to note that these ideas are qualitatively in agreement with experiment. For example, measurements of the rate of nucleation of pearlite shows that nucleation continues for a substantial fraction of the transformation; whilst the kinetics of the precipitation of Fe_3C from ferrite, which is continuous, are consistent with all nuclei being formed within a negligible time interval at the start of reaction.

(iii) Reversion

The theory predicts that the critical size of the nucleus decreases rapidly with temperature as shown in Fig. 5.1. Thus, particles of product that are stable at low temperature may be unstable at a higher temperature. Suppose a specimen is quenched to temperature

---

† J. B. Zeldovich, *Acta. Physicochim. U.R.S.S.* **18**, 1 (1943); D. Turnbull, *Trans. A.I.M.E.* **175**, 774 (1948).

T_3 at which the critical nucleus radius is r_{c3}, allowed to partly transform during which growth to a radius r_a occurs, and then up-quenched to temperature $T(T_E > T > T_3)$ at which the critical radius is r_c. If r_a is larger than r_c growth continues and transformation proceeds normally, but if r_a is smaller than r_c the particles of product formed at T_3 are unstable at T and redissolve even though the product is the thermodynamically stable phase. Reaction re-starts after a further incubation period. This phenomenon, termed *reversion* or *retrogression*, has been noted in several precipitation reactions.

5.7. The Classical Theory of Homogeneous Nucleation in Two-component Systems

An extension of the classical theory of nucleation to transformation in two-component systems was made by Becker.† He

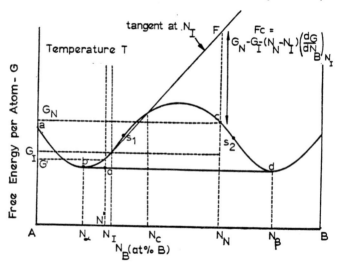

FIG. 5.9a. Free energy composition curve at a temperature T for an alloy system whose equilibrium diagram is as in 5.9b. Consult the text for a more detailed explanation.

† R. Becker, *Ann. Phys.* **32**, 128 (1938); *Proc. Phys. Soc.* **52**, 71 (1940).

NUCLEATION

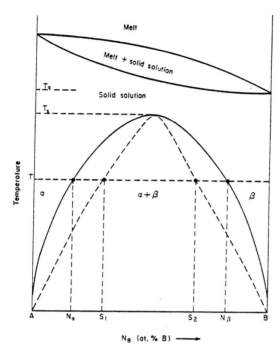

Fig. 5.9b

treated the case of a transformation in which only composition changes are involved. This corresponds to the precipitation from a supersaturated solution of a precipitate having the same structure as the solution but a different composition. Effects due to strain energy are neglected, a procedure that is justified if the parent is either a liquid or a solid at high temperature and thus unable to support a stress; or alternatively if the two solid phases involved have an identical specific volume (lattice parameter) so that no change in volume accompanies the phase change.

Embryos arising as the result of thermal fluctuations in a two-component system vary in shape, size, composition and internal chemical uniformity. The volume free energy change accompanying the formation of an embryo depends upon the composition of both the alloy and embryo. These relations are best analysed by reference to a free energy–composition diagram. The type of diagram for a system satisfying the requirements of the model is shown in Fig. 5.9a for a temperature at which two solutions co-exist. In this G the free energy per atom is given as a function of N_B the atom fraction of component B for a continuous series of solutions of one crystal structure. In this case G is a continuous function of N_B. Applying the criterion of minimum free energy it follows that for alloys in the range pure A to N_α, and N_β to pure B, the solution is the stable form. In the range N_α–N_β, the free energy is minimized by the co-existence of two phases of composition N_α and N_β; N_α and N_β are the compositions at which the common tangent touches the free energy curve for the solution. As the temperature is raised the central hump in the free energy diagram becomes flatter and the composition of the co-existing phases approach each other. In appropriate cases the hump disappears completely and the miscibility gap closes. At higher temperatures all alloys are single phase. The corresponding phase diagram is shown in Fig. 5.9b. In many systems melting occurs before complete mixing and the result is a eutectic with terminal solid solutions.

The two points S_1 and S_2 in Fig. 5.9a are points of inflection (i.e. points at which $\partial^2 G/\partial N_B^2 = 0$) and are called *spinodal* points. The locus of the spinodals with temperature is shown by the broken lines in the phase diagram.

When an alloy, composition N_I is quenched from a temperature T_Q in the single phase region to a lower temperature T at which the free energy-composition curve is that in Fig. 5.9a the initial free energy is that for the solution, i.e. the point G_I. The free energy after complete transformation is a on the common tangent. The first step in applying the nucleation theory is to calculate the change in free energy produced by the formation of an embryo. To maintain generality the embryo composition is regarded as a variable. Let an embryo have a uniform composition N_N. From the curve the free energy corresponding to this composition is G_N. The formation of this embryo causes the matrix to change in

NUCLEATION

composition to N^1, corresponding free energy G^1. The lever rule gives the fraction of N_N as $(N_I - N^1)/N_N - N^1)$ and that of N^1 as $(N_N - N_I)/(N_N - N^1)$. The change in free energy per unit volume of precipitate is

$$\Delta G_v = \frac{1}{\lambda}\left[\left(\frac{N_I - N^1}{N_N - N^1}\right)G_N + \left(\frac{N_N - N_I}{N_N - N^1}\right)G^1 - G_I\right]$$
$$\left(\frac{N_N - N^1}{N_I - N^1}\right) \quad (5.27)$$

where λ is the mean volume per atom in the precipitate.

Rearranging eqn. (5.27) gives

$$\Delta G_v = \frac{1}{\lambda}\left[G_N - G^1 - (N_N - N^1)\left(\frac{G_I - G^1}{N_I - N^1}\right)\right] \quad (5.28)$$

Since the formation of one embryo has little effect on the average matrix composition N^1 and G^1 differ insignificantly from N_I and G_I so that

$$\left(\frac{G_I - G^1}{N_I - N^1}\right) \simeq \left(\frac{dG}{dN_B}\right)_{N_I}$$

Therefore

$$\Delta G_v = \frac{1}{\lambda}\left[G_N - G_I - (N_N - N_I)\left(\frac{dG}{dN_B}\right)_{N_I}\right] \quad (5.29)$$

$(dG/dN_B)_{N_I}$ is the gradient of the G–N_B curve at the initial composition. It follows from simple geometry that

$$\Delta G_v = \frac{1}{\lambda}[FC] \quad (5.30)$$

where FC is the vertical distance, measured at the embryo composition, between the G–N_B curve and the tangent drawn at the alloy composition. Equations (5.29) and (5.30) permit evaluation of ΔG_v from a free energy composition curve. In practice the G–N_B curve is either calculated theoretically on the basis of some model of the system or derived empirically from the phase diagram.

It follows from eqns. (5.29) and (5.30), and Fig. 5.9 that all embryos of composition between N_C and N_β have a negative value of ΔG_v and are stable provided that their size is not less than the critical size determined by the surface energy. Thus, the initial nucleus for precipitation may have any composition in this range. The embryos produced in most abundance are those for which W is a minimum. W is determined by a balance between ΔG_v and γ. Although embryos of the stable phase N_β have the maximum (negative) value of ΔG_v, it does not necessarily follow that they have minimum W. It may happen that an alternative composition has a much smaller interfacial energy, sufficient to compensate for the reduced ΔG_v. In this case a metastable transition phase will be nucleated in preference to the stable phase because this process leads to a more rapid reduction in free energy. The question of transition phases is discussed more fully in section 5.13.

Before further progress can be made with the theory it is necessary to introduce some arbitrary assumption about the composition of the nucleus. In Becker's extension of the classical theory it is assumed that the nucleus has the same composition as the stable phase N_β, that it is internally uniform and that it is bounded by a definite interface where the composition changes discontinuously, from N_β to that of the matrix. Becker used these assumptions as the basis of an analysis of nucleation in a metastable regular solution. The free energy of mixing per atom ΔG^M for a regular solution of composition N_B is

$$\Delta G^M = U_E N_B(1 - N_B)z + kT[N_B \ln N_B + (1 - N_B) \ln (1 - N_B)] \quad (5.31)$$

z is the co-ordination number of the lattice and U_E is the exchange interaction energy given by

$$U_E = [U_{AB} - \tfrac{1}{2}(U_{AA} + U_{BB})] \quad (5.32)$$

where U_{AA}, U_{BB} and U_{AB} are the energies of the bonds between two neighbouring A atoms, two B atoms and one A and one B atom respectively. Application of eqn. (5.31) to each term in eqn. (5.29) with N_N placed equal to N_β gives ΔG_v in terms of U_E.

The problem of evaluating γ was solved by assuming the lattices of nucleus and matrix to be in perfect registry. The energy of the interface is then the sum of the nearest neighbour bonds across that plane at which the composition changes abruptly. Becker showed that

$$\gamma = 2U_E z(N_\beta - N_I)^2 . n_s \qquad (5.33)$$

where n_s is the number of atoms in unit area of the surface of the nucleus. The value of U_E is related to the temperature at which the misibility gap closes (T_s in Fig. 5.9). From the theory of regular solutions T_s is given by

$$zU_E = 2kT_s \qquad (5.34)$$

Combination of eqns. (5.29), (5.31), (5.33), (5.34) with (5.3) and (5.4) gives the size of the critical nucleus and the energy of formation as a function of temperature for an assumed shape.

Becker applied this treatment to precipitation in gold–platinum alloys. This system is completely misible at high temperatures, the solution being f.c.c. ($z = 12$), but separates into two f.c.c. phases at low temperature. The temperature of critical mixing is 1500°K. However, the system is only approximately regular as shown by the asymmetry of the phase diagram. It was assumed that the activation energy for the movement of the interface, U_I in eqn. (5.10), is identical with that for diffusion in the alloy considered, i.e. 39K. cal/mole for a 30 At % Pt alloy. Application of the analysis in 5·3 gave 900°K as the temperature of maximum nucleation rate with a critical nucleus size containing in the order of 100 atoms. Experimentally it is found that the time to half transformation is a minimum at 860°K in good agreement with the theory.

Although this agreement is encouraging too much significance should not be placed upon it. Indeed, it is probably fortuitous in view of the neglect of any strain energies, the arbitrary assumption concerning the composition of the nucleus and the nature of the interface and the fact that half reaction times are only vaguely related to the steady state nucleation rate. It is preferable to regard this agreement as merely justifying the use of the theory as a qualitative model for discussing solid state phase changes.

In two-component systems the other important variable besides temperature is the initial composition. The variation of ΔG_v with initial composition for a fixed nucleus composition is given by differentiating eqn. (5.29) with respect to N_I

$$\frac{\partial(\Delta G_v)}{\partial N_I} = \frac{1}{\lambda}[N_I - N_N]\left(\frac{\partial^2 G}{\partial N_B^2}\right)_{N_I} \quad (5.35)$$

Thus ΔG_v has a maximum value when $(\partial^2 G/\partial N_B^2) = 0$, i.e. when the alloy composition coincides with the spinodal point. It follows that provided the composition of the nucleus does not vary the value of W attains a minimum value at the spinodal; or conversely that the nucleation rate is a maximum. There are two ways of testing this result. The first is to study the rate of nucleation at constant temperature in a series of alloys. The second and easier way is to study one alloy over a range of temperatures, including that at which the composition of the alloy coincides with the spinodal. In this case the theory predicts that the reaction rate should be a maximum at the spinodal temperature. This is consistent with the observation that the nose of the C-curve for precipitation coincides with the spinodal in several systems.

5.8. The Borelius Theory of Nucleation in Two-Component Systems[†]

The Becker treatment assumes that the composition of the embryos is constant and that they become nuclei by fluctuations in size; the basic assumption of the Borelius model is that a group of atoms of fixed size achieve stability by fluctuations in compo-

---

[†] G. Borelius, *Ann. Phys.* **28**, 507 (1937); *Arkiv Mat. Astron. Fysik.* **32A**, 1 (1944); *Trans .A.I.M.E.* **191**, 477 (1951).

sition. The type of system assumed is similar to that in the Becker treatment in that strain energies are neglected. In addition interfacial energies are also ignored. Consequently ΔG equals ΔG_v times the volume of the embryo.

The analysis of this model is based upon equations analogous to eqns. (5.29) and (5.35). The free energy change accompanying the formation of an embryo of composition N_N and containing i atoms is

$$\Delta G = \left[G_N - G_I - (N_N - N_I) \left(\frac{\partial G}{\partial N_B} \right)_{N_I} \right] i \qquad (5.36)$$

As before the right-hand side of eqn. (5.36) is the vertical distance at the composition of the embryo from the tangent at N_I to the G–N_B curve (Fig. 5.9). The value of ΔG for various fluctuations is shown graphically as curve (a) in Fig. 5.10. It is zero when $N_N = N_I$ and $N_N = N_c$; it has a minimum at $N_N = N_\beta$ and passes through a maximum at $N_N = N^*$. The value of N^* can be found by differentiating eqn. (5.36) with respect to N_N keeping N_I and i constant, and putting the derivative equal to zero.

$$\frac{\partial(\Delta G)}{\partial N_N} = \left[\left(\frac{\partial G}{\partial N_B} \right)_{N_N} - \left(\frac{\partial G}{\partial N_B} \right)_{N_I} \right] i = 0 \qquad (5.37)$$

Hence N^* is that composition at which $(\partial G/\partial N_B) = (\partial G/\partial N_B)_{N_I}$, i.e. that composition at which the tangent to the G–N_B curve is parallel to the tangent at the composition of the initial alloy. This is illustrated in Fig. 5.11.

Fluctuations giving an embryo with a concentration of B atoms less than N^* are unstable; clusters with concentration greater than N^* can further concentrate towards N_β because this reduces the free energy. The free energy barrier to the formation of a cluster of critical composition N^* is $i\Delta G^*$ and so the steady state rate of nucleation is given by

$$I = A\, e^{-(i\Delta G^* + U_I)/kT} \qquad (5.38)$$

and the time to form a nucleus

$$\ln t = \text{const} + \frac{i\Delta G^*}{kT} + \frac{U_I}{kT} \qquad (5.39)$$

where U_I has the same significance as before, and ΔG^* the value of ΔG at N^*, can be determined if the free energy–composition curve is known. The difficulty with this theory is that there is no way of estimating i and it has to be regarded as an adjustable parameter, to be determined experimentally.

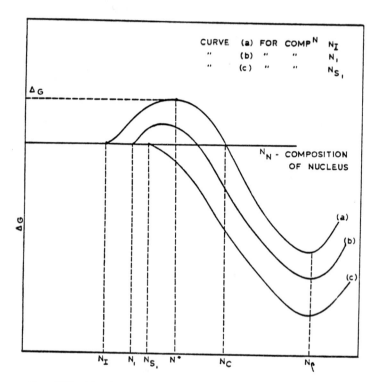

Fig. 5.10. The free energy change accompanying fluctuations in composition, according to Borelius.

As the alloy composition is increased towards the spinodal the value of ΔG^* decreases becoming zero when N_I coincides with the spinodal. This becomes evident when it is noted that as N_I increases the slope of the tangent drawn at N_I increases and so the composition at which the parallel tangent is drawn decreases and the distance between the two tangents also decreases. At the point of inflection the two coincide and thus ΔG^* becomes zero. Curves (b) and (c) in Fig. 5.10 illustrate this effect. The physical interpretation of this result is that for all alloys within the spinodals (i.e. within the range $(\partial^2 G/\partial N_B^2) < 0$) there is no

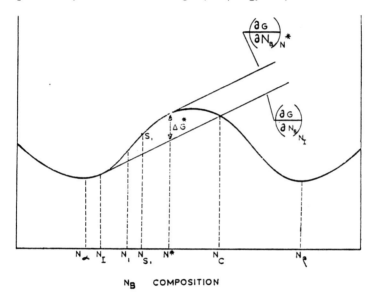

FIG. 5.11. Shows the position of the important compositions in Fig. 10 particularly the critical composition N^* at which the tangent to the G–N_B curve is parallel to the tangent at the alloy composition N_I.

thermodynamic barrier to nucleation and all fluctuations, irrespective of composition, are stable. In view of eqns. (5.38) and (5.39) it follows that there is a noticeable retardation in nucleation rate for a given alloy as the temperature of reaction is increased through the spinodal, because ΔG^* changes from zero to a finite value at that temperature.

To test this theory Borelius and his associates fitted free energy–composition curves to the solubility lines of a number of binary systems which undergo precipitation from supersaturated solutions at low temperatures. Once the $G - N_B$ curve is established the value of ΔG^* can be derived from eqn. (5.36) and the spinodal lines plotted by applying the condition $(\partial^2 G/\partial N_B^2) = 0$. The "temperature of retardation" was determined by measuring the reaction rate over a series of temperatures, assembling the

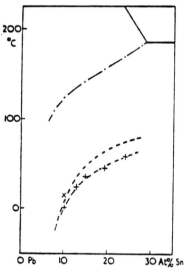

Fig. 5.12. Solid solubility curve and calculated spinodal curve for Pb–Sn alloys. Crosses are temperatures of retardation in precipitation experiments. (After Borelius.)

results into a *TTT* diagram using a reciprocal temperature scale and finding the point at which the *C*-curve deviates from the low temperature linear part. This procedure is identical with that shown in Fig. 5.8. The calculated spinodal line for the lead–tin system with the retardation temperatures for a number of alloys are shown in Fig. 5.12. The agreement is good.

In addition to this, use of the construction described in Fig. 5.8 gives values of $(i\Delta G^*)$ at each temperature, enabling i to be

NUCLEATION

estimated. The value of i is 100–200 atoms depending on temperature and composition.

A serious fundamental difficulty with this theory is that it incorrectly predicts the temperature dependence of the nucleation rate at temperatures near to T_E. For any alloy, T_E is that temperature at which the composition coincides with the composition of

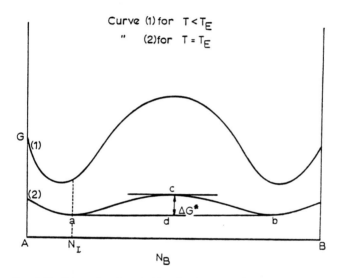

FIG. 5.13. Illustrating that the Borelius theory predicts a finite barrier to nucleation at T_E. An alloy N_B is supersaturated at temperature T, curve (1) and just saturated at T_E, curve (2).

one of the minima, as shown in Fig. 5.13. Thus at T_E the tangent at the alloy composition is the horizontal line ab and the nucleation barrier ΔG^* is given by

$$\Delta G^* = cd \qquad (5.40)$$

It follows that the barrier to nucleation is *finite* at T_E implying a finite rate of nucleation at all temperatures up to and including T_E. In practice, nucleation vanishes at T_E, a finite undercooling being required to start a transformation, showing that the nucleation barrier is infinite at T_E.

5.9. The Hobstetter–Scheil Model for Nucleation in Two-Component Systems

Hobstetter,† and later Scheil,‡ attempted to combine the Becker and Borelius theories. They assumed that the most probable nucleus is that combination of size and composition requiring the least free energy of formation taking into account both surface energy γ and the volume free energy ΔG_v.

The free energy change accompanying the formation of an embryo containing i atoms and of composition N_N in a regular solution is

$$\Delta G = i\left[G_N - G_I - (N_N - N_I)\left(\frac{\partial G}{\partial N_B}\right)_{N_I} \right] + \alpha z(N_N - N_I)^2 \tag{5.41}$$

The first term is ΔG_v for an embryo of variable size i and variable composition N_N, and the second is the surface energy term obtained by replacing N_β by the variable N_N in Becker's evaluation of γ in a regular solution (p. 122). α is a geometric factor, which strictly is also a variable but which is kept constant by assuming a nucleus shape.

ΔG was obtained as a function of i and N_N and assembled into a free energy surface. This surface contains a saddle point. ΔG at this point is the free energy of formation of a stable nucleus. W and r_c and N_c^* are the critical size and composition respectively. The result differs from the Borelius theory in that the concentration of B atoms in the nucleus is never less than N_c. r_c is not the same as in Becker's treatment because the composition of the nucleus is no longer the stable composition N_β. This result is to be expected from Becker's evaluation of γ (eqn. 5.33). γ is reduced by reducing the composition of the nucleus, but this also causes a reduction in ΔG_v. The actual value of N_N represents a balance between these two terms. At temperatures (or compositions) approaching the equilibrium (i.e. at low supersaturation) the

---
† J. N. Hobstetter, *Trans. A.I.M.E.* **180**, 121 (1949).
‡ E. Scheil, *Z. Metallk.* **43**, 40 (1952).

composition of the nucleus approaches N_β. At high supersaturation the nucleus composition is quite different from N_β.

5.10. The Cahn and Hilliard Non-uniform Model of Nucleation in Two-component Systems†

A serious objection to the classical theory and the Hobstetter modification to it is the implied assumption that a sharp interface exists between nucleus and matrix, at which the composition and/or structure change abruptly, and which has a unique value of the interfacial energy independent of curvature. Cahn and Hilliard attempted to overcome this difficulty by regarding the size, composition and internal uniformity as variables, the only assumption being that the nucleus is spherical. The free energy per unit volume G_v of an isotropic, incompressible two-component fluid containing heterogeneites in composition is the sum of two terms. The first is the free energy that the volume has in the absence of non-uniformities and the second a gradient energy which is a function of the local variation in composition. Thus

$$G_v = \int [G(N_B) + \kappa(\nabla N_B)^2] \, dV \tag{5.42}$$

where $G(N_B)$ is the free energy per unit volume in a homogeneous solution of composition N_B and $\kappa(\nabla N_B)^2$ is the term that arises because of composition gradients. This relation holds for any kind of composition non-uniformity and can thus be applied to either local fluctuations in composition or to the interface between two phases of different composition. Furthermore, eqn. (5.42) may be applied to solid state systems provided that the two components have the same atomic size and crystal structure. These restrictions are necessary to ensure that the free energy is a continuous function of composition at constant temperature (i.e. the free energy–composition curve is of the type in Fig. 5.9a) and that variations in composition do not produce any

---

† J. W. Cahn and J. E. Hilliard, *J. of Chem. Physics*, **28**, 258 (1958); **31**, 688 (1959); *Acta. Met.* **9**, 795 (1961); **10**, 179 (1962).

strain in the lattice. These assumptions are the same as those made in the Becker and Borelius theories in sections 5.6 and 5.7.

Nucleation in a metastable solution is that fluctuation in composition corresponding to the saddle point on the free energy surface. By applying eqn. (5.42) Cahn and Hilliard showed that at low supersaturation (i.e. as $T \to T_E$ or $N_I \to N_a$) the nucleus has a uniform composition N_β and a sharp interface, the energy of which does not vary with curvature. As $T \to T_E$ the radius of the nucleus and the activation free energy of formation tend to infinity. These conclusions are in accord with the classical model. As supersaturation increases the work of forming a critical nucleus becomes progressively less than that predicted by the classical theory, eventually becoming zero at the spinodal. As the spinodal is approached the radius increases to infinity and the interface becomes more and more diffuse until eventually no part of the nucleus is even approximately uniform. For an alloy decomposing under these conditions the concept of an interface ceases to have physical reality.

This theory is a considerable improvement over previous treatments but mathematically it is much more complicated and a detailed discussion is beyond the scope of this book. It seems clear that the theory supports, at least qualitatively, the predictions of the classical treatment at high temperatures and low supersaturations. At low temperatures the predictions are more in accord with the Borelius fluctuation theory insofar as it shows that there is no energy barrier to the formation of a stable nucleus for all compositions within the spinodal lines. In subsequent sections the simple concepts and formalism of the classical theory are used as the basis of the discussion. But the limitations of this treatment demonstrated by Cahn and Hilliard should be borne in mind.

5.11. The Influence of Strain on Nucleation in the Solid State

The treatment of nucleation in the foregoing section is applicable strictly only to those phase changes in which the parent and product have identical specific volumes. In the majority of phase transformations a volume change occurs due to either a change in structure or lattice parameter. The volume change is occommodated by elastic strain in either or both of the two phases and it is only permissible to neglect it when one of the phases is incapable of supporting a stress, e.g. when one of the phases is a liquid or when the alloy is solid but at high temperature. In addition the tendency for crystalline solids to form coherent or partially coherent interfaces results in lattice strains due to disregistry across the interface.

In writing an expression for the free energy change accompanying the formation of an embryo, it is necessary at this stage to drop the assumption of a spherical habit because this is unlikely when strain energy is important. If σ is the increase in strain energy accompanying the formation of unit volume of embryo, the change in free energy is

$$\Delta G = i(\Delta G_v + \sigma)\lambda + \alpha i^{2/3}\gamma \qquad (5.43)$$

where i is the number of atoms in the embryo, λ is the atomic volume in the embryo and α is a factor depending upon the shape of the embryo. σ is always positive and thus increases the size of the critical nucleus and the activation energy for its formation. The number of atoms in a critical nucleus i_c and the free energy of formation W can be found by the same procedure as in section 5.3 for any assumed nucleus shape.

Incoherent nucleation. An incoherent interface is one across which there is no crystallographic continuity. The structure and/or composition change rapidly from that characteristic of one phase to that of the adjoining phase over a distance of a few atoms. The value of γ is fairly insensitive to composition and orientation difference and is of the order of 200–1000 erg/cm².

In the absence of lattice continuity the strain associated with the formation of an embryo in a crystalline lattice is that associated with the hydrostatic strain produced by placing an inclusion of volume V into a hole of volume $(1 + \Delta) V$, where Δ can be either positive or negative. ΔV, the transformation dilatation, is accommodated by either plastic flow if α and β are sufficiently soft, or by elastic strain in α and β when the phases are rigid. In the former case the strain energy is negligible; in the second σ depends upon the values of the elastic modulii and of the shape of the particle, as shown by Nabarro.†

Nabarro considered the embryo to be in the form of a spheroid, of semi-axes, a, a and c. When $c/a = 1$ the shape is a sphere;

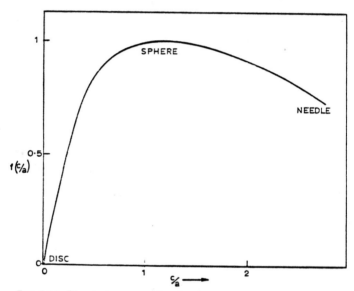

FIG. 5.14. The strain energy of an incoherent spheroidal nucleus as a function of its shape. a is the radius and $2c$ the thickness of the spheroid. (After Nabarro.)

$c/a \gg 1$ corresponds to a needle; $c/a \ll 1$ approximates to a thin disc. On the assumption that the strain is accommodated predominantly in the matrix, the strain energy per unit volume of particle σ is given by

$$\sigma = \tfrac{2}{3}\mu(\Delta)^2 f(c/a) \qquad (5.44)$$

μ is the modulus of rigidity of the matrix.

---

† F. R. N. Nabarro, *Proc. Roy. Soc.* **A175**, 519 (1940); *ibid.* **52**, 90 (1940).

If the formation of the embryo involves a one-for-one interchange of atoms Δ is simply the fractional difference between the volume per atom in the embryo and the matrix. However, it is to be expected that the stress field around the embryo will induce a flow of vacancies to or from the interface to minimize σ by effectively reducing Δ. The function $f(c/a)$ has a maximum value of unity at $c/a = 1$ (sphere), a value of 0·75 when $c/a = \infty$ (cylinder) and is approximately equal to $\frac{3}{4}\pi c/a$ when $c/a \ll 1$. $f(c/a)$ could not be evaluated for other values of c/a but on the reasonable assumption that it varies smoothly and continuously it must be as shown in Fig. 5.14. It is clear that the strain energy is minimized when the precipitate is in the form of thin plates.

Although an incoherent embryo in the form of a thin plate has a minimum strain energy it has a high surface area–volume ratio and so is unfavourable from the point of view of surface energy. The actual shape is an oblate spheroid being a compromise between the two opposing factors. Noting that the volume of an oblate spheroid is

$$\text{volume} = \tfrac{4}{3}\pi a^2 c \qquad (5.45)$$

and the surface area is

$$\text{area} = \pi a^2 \left[2 + \frac{c^2}{a^2 \cdot e} \cdot \ln\left(\frac{1+e}{1-e}\right) \right] \qquad (5.46)$$

in which e, the eccentricity, is $e = (1 - (c^2/a^2))^{1/2}$ the free energy of formation of an embryo is given by

$$\Delta G = \tfrac{4}{3}\pi a^2 c [\Delta G_v + \tfrac{2}{3}\mu(\Delta)^2 f(c/a)] +$$
$$\pi a^2 c \left[2 + \frac{c^2}{a^2 e} \ln\left(\frac{1+e}{1-e}\right) \right] \qquad (5.47)$$

ΔG in eqn. (5.47) represents a surface in four-dimensional space, the independent variables being a, c and the composition of the embryo N_N. The saddle point can be found in principle by finding the value of ΔG which simultaneously satisfies the conditions.

$$\frac{\partial(\Delta G)}{\partial c} = 0 \qquad (5.48)$$

$$\frac{\partial(\Delta G)}{\partial a} = 0 \qquad (5.49)$$

$$\frac{\partial(\Delta G)}{\partial N_N} = 0 \qquad (5.50)$$

A general solution is not possible owing to the uncertainty of $f(c/a)$. An approximate answer may be obtained for disc shaped embryos. Nabarro found that as $c/a \to 0$, $f(c/a) \simeq \frac{1}{2}\pi \cdot c/a$. As c/a tends to zero the second term in the square bracket in eqn. (5.47) becomes negligible in comparison with 2. Carrying out the differentiation, assuming that the composition of the alloy and the nucleus and thus ΔG_v is constant gives

$$c_c = -\frac{2\gamma}{\Delta G_v} \qquad (5.51)$$

$$a_c = \frac{2\pi\mu(\Delta)^2 \gamma}{(\Delta G_v)^2} \qquad (5.52)$$

and

$$W = \frac{8}{3} \frac{\pi^3 \mu^2 (\Delta)^4 \gamma^3}{(\Delta G_v)^4} \qquad (5.53)$$

where a_c and c_c are the equatorial and polar radii of the critical nucleus. ΔG_v may be evaluated for an assumed nucleus composition from eqn. (5.29).

Evidently as the stiffness of the matrix increases the nucleus tends to become thinner to reduce the distortion energy.

Coherent nucleation. When the interface between nucleus and matrix is incoherent the value of γ is large and the interfacial energy term is dominant in determining the formation energy. The energy of coherent or partly coherent interfaces is considerably less than that of the incoherent type and thus a substantial reduction in the free energy of formation of a nucleus results if the embryos are bounded by coherent interfaces.

An interface between two crystals is fully coherent when the plane of atoms constituting the interface, disregarding chemical species, is common to both crystals. A condition for coherency is that the crystals possess crystallographic planes in which the atomic configuration and spacing is identical or very nearly identical. For example, if one crystal is c.p.h. and one f.c.c. with $a_{HEX} = 1/\sqrt{2}\ (a_{cubic})$ then a coherent interface is formed by placing the (0001) plane of the hexagonal crystal in contact with the (111) plane of the cubic crystal such that the $[2\bar{1}\bar{1}0]$ direction in the hexagonal is parallel to $[\bar{1}10]$ in the cubic crystal. Thus, phases joined by coherent interfaces are crystallographically related, i.e. the relative orientation of the crystal is such that specific crystallographic planes in the two crystals are parallel and certain directions in these planes coincide. A description of the orientation relationship requires a statement of the coincident planes and directions. In the example the relationship would be written as

$$(0001)_{HEX}\ ||\ (111)_{cubic}\ \text{and}\ [2\bar{1}\bar{1}0]_{HEX}\ ||\ [\bar{1}10]_{cubic}$$

In addition, if the precipitate habit is plate or rod-like then it is usual to state the habit plane, which is the plane of the matrix parallel to the plane of the plate or the axis of the rod. See, for example, Fig. 5.15a.

In general the atomic spacing in the matching planes is not quite the same. If the spacing in the unstressed matrix is a_α and in the unstressed embryo a_β then the mismatch or disregistry δ is defined as

$$\delta = \left|\frac{a_\alpha - a_\beta}{a_\beta}\right| \tag{5.54}$$

In a perfectly coherent interface the disregistry is accommodated by elastic distortion of the two lattices. This is illustrated schematically in Fig. 5.15b. The extent to which the distortion is distributed between matrix and embryo depends upon the elastic constants of the two crystals measured in the directions contained in the matching planes. If the embryo is much stiffer than the matrix the elastic strain is chiefly in the matrix, and vice versa.

The energy γ_c of a perfectly coherent interface in an alloy may be regarded as the sum of two separate parts: (a) γ_{CH}, a

138 THE KINETICS OF PHASE TRANSFORMATIONS IN METALS

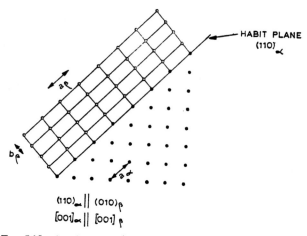

$(110)_\alpha \parallel (010)_\beta$
$[001]_\alpha \parallel [001]_\beta$

Fig. 5.15a. A coherent interface between two phases with perfect registry.

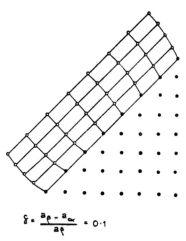

$$\delta = \frac{a_\beta - a_\alpha}{a_\beta} = 0.1$$

Fig. 5.15b. A coherent interface with slight disregistry showing lattice strain.

"chemical term" arising from the short range forces between chemically dissimilar atoms in adjacent positions across the interface; γ_{CH} is the energy estimated by Becker (p. 122) for the interface between two regular solutions, by counting the number of like and unlike bonds; and (b) σ_{SF} a structural term, associated with the long range forces of the lattice distortion produced by the

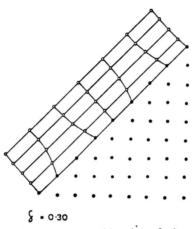

$\delta = 0{\cdot}30$

FIG. 5.15c. A semi coherent interface with regions of coherency separated by dislocations. (Schematic).

disregistry.† σ_{ST} has been calculated for a few simple cases by Nabarro who showed that, in general,

$$\sigma_{ST} = f(c)\delta^2 \qquad (5.55)$$

where σ_{ST} is the lattice strain energy per unit volume of embryo and $f(c)$ is some function of the elastic constants of matrix and embryo and the shape of the embryo. The geometry of the embryo is governed by the magnitude of the strain energy, the relative stiffness of embryo and matrix and the variation of elastic

---

† The symbol σ is used for the component of the coherency energy to emphasize that it depends upon the volume of the precipitate and is thus comparable with the elastic energy σ in equation (5.43).

constants with crystallographic direction in both phases. Nabarro showed that if all the elastic modulii of the embryo are about equal to those of the matrix σ_{ST} is independent of the shape and σ_{ST} is given approximately by

$$\sigma_{ST} = 6\mu\delta^2 \qquad (5.56)$$

If the elastic modulii in one plane in the embryo is less than the modulii in the matrix the preferred shape is that of a thin disc with the plane of the disc in the plane of minimum elastic modulus. However, change in shape reduces the strain energy from that given by eqn. (5.56) by no more than a factor of about six.

For a fixed value of δ the strain energy increases as the size of the embryo increases. Eventually a size is reached at which it is energetically more favourable to form dislocations in the interface in preference to increasing the lattice strain. The interface then consists of regions of complete coherence separated by dislocations (see Fig. 5.15c). This type of boundary is said to be semi-coherent. The spacing of the dislocations is such that their long range stress fields just cancel that due to the disregistry. For example, if $\delta = 0.01$ and the dislocations have a Burgers vector equal to one atomic spacing in the matching plane, a dislocation every $1/0.01 = 100$ spacings just accommodates the mismatch. An interface less than 100 spacings in extent is perfectly coherent; larger ones are semi-coherent. The energy of a semi-coherent boundary depends upon the density of the dislocations contained therein and thus upon δ. By applying the dislocation model of interfaces, Brooks† showed that for a partially coherent boundary

$$\gamma_{ST} = A\delta[B - \ln \delta] \qquad (5.57)$$

where the constants A and B depend upon the elastic constants of the two structures and γ_{ST} is the strain energy *per unit area* of interface.

The values of interfacial energies have been measured in a few systems. The evidence available suggests that the energies of completely incoherent boundaries fall in the range 500–1000

---

† H. Brooks, *Metal Interfaces*, Am. Soc. Metals, 1952.

erg/cm². The value for a coherent interface depends very much on the mismatch, a lower limit being probably about 25 erg/cm² —(the energy of a twin interface in copper)—and the maximum value being about 200 erg/cm². Partly coherent interfaces fall between 200 to 500 erg/cm². Clearly a considerable reduction in surface energy and hence W results when nucleation is coherent.

Complete loss of coherency occurs when the energy of an incoherent interface ($\sim$ 500 erg/cm²) becomes equal to that of the coherency strain. Using eqn. (5.56) the strain energy of a spherical particle is $8\pi r^3 \mu \delta^2$ and the total energy of the boundary after breakaway is $4\pi r^2 \gamma$. Coherence is lost when

$$r = \frac{\gamma}{2\mu\delta^2} \tag{5.58}$$

Taking a value of μ as 4×10^{10} dyn/cm² and $\delta = 0.01$ gives $r \simeq 625$ Å as an approximate size at which coherency is lost. A disc of constant radius is incoherent when the thickness exceeds about 2500 Å. These values are much greater than the dimensions of critical nuclei deduced in previous sections. Consequently, it is to be concluded that solid state nucleation occurs coherently.

Thus the picture that emerges from these ideas is that embryos and nuclei form coherently unless the matrix and embryo structures are so dissimilar that well-matched planes of atoms do not exist. The shape of the coherent or semi-coherent nucleus is governed by the relative values of the elastic constants, the magnitude of the mismatch and the anistropy of the surface energies. As the nuclei increase in size the coherency strains increase until a lowering of the strain energy is realized by the partial breakdown of coherency. The number of dislocations in the interface increases as the size increases until eventually the dislocation density is so great that all coherency is lost. At this stage the shear strain in the precipitate has vanished and it remains under a hydrostatic pressure described in the previous section.

Widmanstätten structures. Many phase changes in the solid state produce structures in which the crystal of the product phase have a plate or needle habit and are arranged parallel to certain crystallographic planes of the surrounding matrix. The resulting microstructure is termed a Widmanstätten structure. The formation of the patterns is readily understood in terms of the ideas in the previous section. Nucleation occurs coherently causing an orientation relationship to be established and the habit plane of the matrix to be selected. After breakaway all the elastic strain is due to the hydrostatic pressure on the growing particles and it follows from 5.11 that there is a great tendency for the particles to develop a plate or rod-like form, if this does not exist already, to minimize strain energy.

5.12. Heterogeneous Nucleation in the Solid State

There is abundant evidence that phase changes of all kinds are nucleated heterogeneously at structural imperfections. Nucleation at free surfaces and grain boundaries is easily observed with the optical microscope. Electron microscopy has confirmed that dislocations and stacking faults are also effective sites. The work of forming a critical nucleus is less at structural imperfections than in a perfect lattice. A reduction in W may arise by either (a) a reduction in the magnitude of either or both γ and σ and/or (b) the contribution of a negative term to the free energy of formation of embryos as a result of the disappearance of the defect and the release of its free energy. It is convenient to examine each type of imperfection individually.

Nucleation at foreign surfaces. This has already been examined in relation to solidification. Analogous arguments apply to the solid state. Nucleation at free surface is encouraged because of the ease of accommodating transformation dilatations.

Grain boundaries. When a nucleus forms at a grain boundary a small part of the boundary disappears and the grain boundary energy released reduces the energy of forming the nucleus. Suppose that a nucleus of β in the form of an oblate spheroid, of equatorial radius a, and polar radius c, forms on a grain boundary of α. The surface area of the nucleus is approximately $2\pi a^2$

NUCLEATION 143

(when $a \gg c$) and the area of boundary destroyed is πa^2. Neglecting strain energy, the free energy of formation is

$$\Delta G = \tfrac{4}{3}\pi a^2 c\Delta G_v + 2\pi a^2 \gamma - \pi a^2 \gamma_{GB} \qquad (5.59)$$

where γ is the energy of the α–β interface assumed incoherent and γ_{GB} is the grain boundary energy. In the limit of $\gamma_{GB} = 2\gamma$ the barrier to nucleation disappears. This is unlikely ever to happen because grain boundary energies are roughly the same as incoherent interfaces. Taking $\gamma_{GB} = \gamma$ reduces the surface energy term by one half and W by one-eighth.

Still further reductions in W is accomplished at junctions between 3 grains or 4 grains. For example, Clemm and Fisher† showed that the activation energy for nucleation at the latter is about $1/2000$ that of homogeneous nucleation.

Other reasons for preferred nucleation at grain boundaries are

(*a*) certain types of solute atoms segregate to grain boundaries, which facilitate the process of assembling sufficient atoms to form the nucleus.

(*b*) Diffusion is more rapid along boundaries.

Dislocations. It is well established that dislocations are preferred sites for precipitation from solid solutions. A simple model of dislocation nucleation was discussed by Cahn,‡ assuming an elastic model of a dislocation and an incoherent interphase interface. According to Cahn the free energy of formation of a nucleus consists of three terms; the volume free energy term, the surface energy term of similar characteristics to the homogeneous theory, and a strain–energy term which is *negative* on account of the release of the strain energy of the dislocation. The value of ΔG for a cylindrical nucleus of unit length and radius r is

$$\Delta G = -A\log r + 2\pi.\gamma.r + \pi r^2 \Delta G_v \qquad (5.60)$$

where A is given by dislocation theory in terms of the elastic constants.

If $|2A.\Delta G_v| < \pi\gamma^2$, ΔG passes through a minimum at a radius

---

† P. J. Clemm and J. C. Fisher, *Acta Met.* 3, 70 (1955).
‡ J. W. Cahn, *Acta Met.* 5, 169 (1957).

r_0, and a maximum at r_c the critical nucleus as shown in curve A in Fig. 5.16. r_0 is roughly equivalent to the size of a Cottrell atmosphere and is given by

$$r_0 = -\frac{\gamma}{2\Delta G_v}\left[1 - \sqrt{1 + \frac{2A\Delta G_v}{\pi \gamma^2}}\right] \quad (5.61)$$

The nucleation barrier is the difference between ΔG at r_c and at r_0. When $|2A.\Delta G_v| > \pi\gamma^2$ there is neither minimum nor maxi-

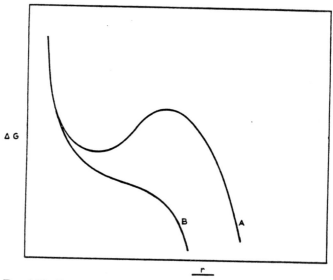

FIG. 5.16. Free energy of formation of a cylindrical nucleus of unit length and radius r along a dislocation. Curve A is for the case where $|2A.\Delta G_v| < \pi\gamma^2$ and B for the case where $|2A.\Delta G_v| > \pi\gamma^2$. (After Cahn.)

mum and consequently no barrier to nucleation, curve B. Taking reasonable values for the parameters, r_0 works out as about 2 Å and r_c, 10 Å. For typical conditions Cahn showed that dislocation nucleation is 10^{78} faster than homogeneous nucleation.

The time dependence of heterogeneous nucleation. The theory of heterogeneous nucleation predicts that several types of nucleating sites are present in a specimen, each type characterized by a particular value of W. Experience with the solidification of small droplets, discussed in section 5.4, confirms that this is so. Growth of nuclei is a much easier and more rapid process than the formation of the nuclei and thus it is to be expected that in bulk samples, transformation is accomplished from those sites having the minimum value of W, unless the number of these sites is very small. Since the number of these sites is very much smaller than the total number of atoms in the crystal, the rate of nucleation must decrease with time as the sites are exhausted or rendered ineffective by being included in transformed regions. This problem was treated by Avrami[†] for the case of discontinuous reactions. Let the number of possible sites per unit volume of parent phase be N_0. This remains constant during a discontinuous reaction because the matrix is unaffected until swept over by the parent–product interface. The number of sites ingested by the new phase in time dt is $N_0.dV$ where dV is the increase in the volume of product phase in the same time.

The probability of a site becoming active is $Ae^{-(W+U_I)/kT}$, A being a frequency factor, W the work of forming a nucleus at the site and U_I the activation energy for interface movement. If $N(t)$ is the number of potential sites available per unit volume of matrix at time t, then the decrease in time dt is

$$- Ae^{-(W+U_I)/kT}.N(t)dt \tag{5.62}$$

The total number dN lost is thus given by

$$dN = - N_I.dV - Ae^{-(W+U_I)/kT}.N(t).dt \tag{5.63}$$

Two limiting cases are of interest

(a) N very large so that loss by ingestion is negligible. Then

$$\frac{dN}{dt} = - Ae^{-(W+U_I)/kT} N(t) \tag{5.64}$$

Integrating,

$$N(t) = N_0 e^{-Ae^{(-(W+U_I)/kT)t}} \tag{5.65}$$

---

† M. Avrami, *J. Chem. Physics*, **7**, 1103 (1939); **8**, 212 (1940).

This expression now replaces the constant N in the equation for the rate of nucleation, i.e.

$$I = N_0 e^{-Ae^{(-(W+U_I)/kT)t}} A e^{-(W+U_I)/kT} \qquad (5.66)$$

It follows that the rate of nucleation decreases exponentially with time at constant temperature.

(b) N very small so that ingestion is the predominant effect

$$\frac{dN}{dt} = N_0 \frac{dV}{dt} \qquad (5.67)$$

which is equivalent to a constant rate of nucleation per unit transformed volume.

In the case of continuous diffusional reactions a growing particle draws atoms from a considerable volume of matrix causing a decrease in the driving force for nucleation ΔG_v throughout this volume. It follows that nucleation is unlikely in these regions. It will be shown in the next chapter that in the case of dilute solution, diffusional growth leads to a uniform decrease in concentration throughout the specimen. The implication is that nucleation proceeds for only a very short time at the start of the reaction and then stops completely.

In addition to the time dependence due to site exhaustion, the transients in the formation of the steady state embryo distribution has to be considered. As noted in section 5.6 this predicts an initial acceleration of the nucleation rate. The actual variation with time in particular cases will depend upon the balance between the two opposing effects.

5.13. The Formation of Metastable Transition Phases

It was pointed out in section 5.6 in connection with Fig. 5.9 that for phase changes in alloys a wide range of nuclei compositions have a negative ΔG_v and thus are thermodynamically possible. The first product of the decomposition of a metastable system is not necessarily the most stable phase, even though it is associated with maximum driving force. A metastable product of lower stability will form preferentially if it can be nucleated at a significantly greater rate, i.e. if its value of W is less. Since ΔG_v is less for a transition phase than for the most stable phase the former can form only if its surface energy is substantially less than that associated with the stable state. Systems in which this is likely are

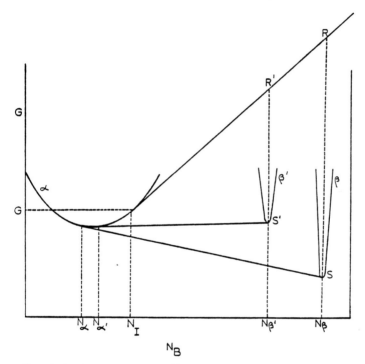

FIG. 5.17. Free energy composition curves for a system forming a transition phase, β^1, prior to the stable phase β.

those in which the stable precipitate β is very unlike the matrix α in structure and composition so that any form of coherency is not possible. A metastable product with a structure and composition such as to facilitate coherent nucleation will have an appreciably lower value of γ and, provided this reduction in interfacial energy more than compensates for the loss in chemical free energy W is less for the transition product. A hypothetical case is illustrated in Fig. 5.17. For an alloy composition N_I the stable state is a mixture of the solution α, of composition N_α, and the phase β. In the supersaturated state its free energy is G. A possible transition phase is β^1 the free energy curve of which lies below the tangent to the α curve at N_I. If the work of forming a β nucleus is W, and that of β^1 is W^1, the condition for β^1 to form in preference to β is $W^1 < W$.

The solubility of β^1 in α is $N_\alpha 1$. Only alloys of composition greater than $N_\alpha 1$ can form the transition phase. Alloys in the range N_α to $N_\alpha 1$ are unsaturated relative to β^1 and these alloys must decompose directly to the stable phase. This illustrates the general rule that as supersaturation decreases, the number of intermediate reaction stages decrease.

Finally, it is necessary to examine what happens after the formation of the transition phase is complete. The free energy of the system is G^1. Further reduction results from the formation of β, with the result that regions of matrix around these new nuclei are of composition N_α, a value less than the solubility of β^1, and hence the transition precipitate dissolves. This is an example of another general rule: the formation of a more stable product results in the re-solution of less stable phases formed in earlier stages. Many precipitation hardening systems show this behaviour.

5.14. Comparison of the Theory of Solid State Nucleation and Experience

In the theories of nucleation discussed in previous sections embryos are assumed to arise by thermal fluctuations, either within structurally perfect regions of the parent phase—homogeneous nucleation—or at structural imperfections—heterogeneous nucleation. Qualitatively, the theory is in good agreement with experience of phase changes in the solid state. Incubation periods, transformation hysteresis, C-curve *TTT* diagrams, Widmanstätten structures, reversion, precipitation on crystalline defects, and the existence of transition phases are common features correctly predicted by the theory. Only few quantitative comparisons with nucleation in the solid state have been made, and most of these employed experimental data, whose relation to nucleation rates is obscure and thus the results of these tests are of doubtful significance—cf. sections 5.7 and 5.8. Counting the number of domains of product phase is the most direct method of deriving nucleation frequencies, although even this depends upon the domains growing to observable dimensions and thus the data is dependent upon the rate of growth. The technique has been applied to a study of discontinuous precipitation in a Au-38 atomic per cent Ni alloy.† The nose of the C-curve occurs about 130°C below T_E. Sufficient thermodynamic data is available to give ΔG_v without the necessity of assuming a model of a solid solution. Using the results in section 5.1 it is found that if the nucleus is completely incoherent and strain energy is neglected the classical theory value of W is $1\cdot5 \times 10^7$ cal/mole. Taking into account strain energy, estimated from Nabarro's equation (5.44) raises W to 7×10^{12} cal/mole. For a coherent nucleus the calculated value of W is $1\cdot3 \times 10^{10}$ cal/mole. Experimentally W is found to be $2\cdot5 \times 10^4$ cal/mole.

The same type of measurements and calculations have been made for the nucleation of graphite from an austenite–cementite mixture, and the same large discrepancy between calculated and measured value of W found. For example, W for an incoherent nucleus, with strain energy neglected is calculated to be between 1 and 10×10^7 cal/mole depending on the value assumed for γ. The measured value is 7–9×10^4 cal/mole.

In view of the enormous disparity between the experimental and calculated values of W it is clear that the classical theory of nucleation in its present form is not capable of being applied quantitatively to nucleation in the solid state. The failure of this approach may be due to the assumption in the theory that the properties of embryos are the same as those of the bulk phase. This

---

† M. Cohen, *Trans. A.I.M.E.* **212**, 171 (1958).

is extremely unlikely because (*a*) even in a stable phase the energy of small groups of atoms deviates considerably from the mean (cf. section 1.8), (*b*) atomic configurations that minimize the interfacial energy probably exist in the embryos, whereas in the bulk phase, surface effects are negligible and the surface energy plays no part in determining the atomic configuration, (*c*) the interfacial energies derived from bulk samples, generally refer to interfaces that are plane or nearly so, and in using these values for nucleation calculations it is assumed that γ is independent of the radius of curvature, an assumption that is not expected to hold at small radii.

A completely different interpretation that is now being actively explored is that in the solid state nucleation in the classical sense is not involved at all. The situation is analogous to the large discrepancy that exists between the observed critical shear stress of crystals and that calculated for a perfect crystal. Much the same explanation is invoked, i.e. that potential nuclei are always present in a phase, irrespective of whether it is stable or metastable, as groupings of the structural imperfections that are always present. Dehlinger† has emphasized that it is possible to describe a transition from one lattice to a second in terms of a suitable configuration of dislocations. Hence embryos of other lattices exist *permanently* within the stable substructure of a phase. The nucleation of a phase change then requires the increase in size of these embryos to a size at which they become stable. The simplest possible example is the transformation from a f.c.c. to a c.p.h. lattice as in the polymorphic transition in cobalt, which can be nucleated by stacking faults in the f.c.c., which are effectively mono-layers of c.p.h. lattice.

An attractive feature of this approach to the problem is that it permits a unified picture of phase changes. Diffusional changes are those in which the embryos increase in size by means of the climb of the circumscribing dislocations due to the migration of single atoms and vacancies; diffusionless changes are those in

---

† U. Dehlinger, *The Physical Chemistry of Metallic Solutions and Intermetallic Compounds.* Paper 4B. H.M.S.O. (1959).

which growth involves dislocation movement analogous to glide. The model is particularly relevant to the latter group since they often occur at such speeds and temperatures that thermal activation is precluded.

Further Reading for Chapter 5

HOLLOMON, J. H. and TURNBULL, D., *Progress in Metal Physics*, Vol. 4 (Ed. B. Chalmers), Pergamon Press, 1953.

TURNBULL, D., *Solid State Physics*, Vol. 3 (Eds. F. Seitz and D. Turnbull), Academic Press, 1956.

SMOLUCHOWSKI, R., *Phase Transformations in Solids* (Ed. R. Smoluchowski), Wiley, 1951.

HARDY, H. K. and HEAL, T. J., *Progress in Metal Physics*, Vol. 5 (Ed. B. Chalmers), Pergamon Press, 1954.

HOBSTETTER, J. N., Decomposition of Austenite by Diffusional Processes, *Am. Inst. Min. Met. Eng.*, 1963.

KELLY, A. and NICHOLSON, R. B., Precipitation Hardening, *Prog. in Materials Sci.* **10**, 3, Pergamon Press (1963).

POUND, G. M., *Liquid Metals and Solidifications*, Am. Soc. Metals, 1958.

BARRETT, C. S., *Structure of Metals*, 2nd Ed., McGraw-Hill.

TAYLOR, A., *X-Ray Metallography*.

CHAPTER 6

Theory of Diffusional Growth Processes

6.1. Introduction

The actual atom movements involved in the growth of crystalline phases cannot be studied directly but have to be inferred from indirect observations. Studies of orientation relationships between parent and product crystals, and of the constitution, structure and morphology have often provided valuable clues as to the mechanism of growth. But the most powerful method is to correlate the measured rates of growth with those calculated on the basis of an assumed model of the growth process.

Experimentally the study of the kinetics of growth is on a much firmer basis than nucleation kinetics because it is frequently possible to measure rates of growth directly when the volumes of product attain observable dimensions. Two methods are commonly used. The first involves the continuous observation of the dimensions of one particular crystal in a specimen which is transforming on the stage of a microscope. The objection to this method is that the growth of a crystal at the surface is unlikely to be truly representative of one growing in the bulk because transformation strains are more easily accommodated at the surface and because surface diffusion is much more rapid than diffusion through the lattice. The second method, which is more tedious but avoids this difficulty, is to section and examine a series of identical samples previously reacted for various times at the same temperature. It is necessary to assume that the section intercepts a sufficiently large number of growing particles to ensure that the largest observed is an equatorial section through the largest particle present. The largest particle originates from the first formed nucleus, and since nucleation commences at the same time in each specimen, the measurements on each sample are effectively referred to the same time scale. The data from either method are plotted as linear dimensions as a function of time. The slope of the graph gives the rate of growth.

It is to be expected that growth rates in diffusional transformations depend upon some or all of the following factors:

(a) The detailed mechanism by which the interface propagates through the parent lattice. This will vary with the nature of the interface and may possibly be different for different directions in the crystal. For example, movement of coherent or partially coherent interfaces must involve the co-ordinated shift of groups of atoms in order to preserve coherency, whereas an incoherent boundary may migrate through the capture of individual atoms.

(b) The crystallographic relationship between parent and product because the rate of growth of a crystal is different for different crystallographic directions.

(c) The rate of diffusion of the various atoms in both phases since this is the means by which atoms are re-distributed.

(d) The type and concentration of lattice defects in the parent and product.

(e) The variation of solubility with curvature of the interface. Small particles of a phase have a higher solubility than larger ones—see section 6.5—due to the difference in the radii of curvature of the interfaces. Thus the concentration of solute in the matrix in equilibrium with a phase decreases as the particle grows. This effect is negligible for crystals larger than 1μ, but the variation of composition around the surface of a non-spherical particle due to variations in curvature may be important.

(f) The amount of latent heat released and the rate at which it is dissipated.

A full analysis of growth rates taking into account all the variables is a formidable problem and has never been attempted. The usual practice is to assume that the rate of growth is determined by one or possibly two factors and that all others may be

154 THE KINETICS OF PHASE TRANSFORMATIONS IN METALS

neglected. In the sections that follow some simplified treatments along these lines are presented for some cases of particular interest.

6.2. Growth of a Phase in a Single Component System

Examples of interest are the growth of pure metal crystals during solidification or polymorphic transformations. A considerable amount of attention has been devoted to the former but little to the latter.† Much of the thinking about crystal growth has been influenced by work on growth from the vapour phase or from solutions and it is convenient to briefly mention some important

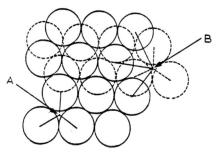

FIG. 6.1. Part section through a step in a close packed plane in a f.c.c. crystal. Full circles are atoms in plane of drawing, broken circles in plane beneath. An atom settling at A makes three bonds (full lines). One settling at a step makes six, three with the atoms beneath (broken lines) and three with adjacent atoms in the same layer as itself (full lines).

points that have arisen from this work. The growth rate of crystals is very sensitive to the crystallographic direction. A freely growing crystal tends to be bounded by planes of high atomic density, i.e. low index planes. This generalization is certainly true of crystals grown from solutions or the vapour; it is reasonably true in the early stages of growth from the melt as evidenced by the crystallographic character of dendrites; it is not true for crystals growing in

---

† See K. A. Jackson, *Liquid Metals and Solidification*, p. 174, Am. Soc. Metals, 1958.

the solid state due to the mutual interference of neighbouring grains. This tendency can be readily understood on the basis of the number of bonds that an atom makes when it encounters the surface of a crystal. If it lands on the surface of a fully-packed (111) plane of a f.c.c. crystal (i.e. point A in Fig. 6.1) only three nearest neighbour bonds are formed, whereas an atom at the edge of a step in this plane, point B, makes on average six bonds, and it is then far more likely to stick than the one at point A.

The initiation of a new layer on a low index plane requires two-dimensional nucleation because the energy of the atoms around the periphery of a disc is higher than that of an atom in the centre of the disc and hence there is a minimum size for stability. The free energy change accompanying the formation of a disc, radius r and thickness t, is

$$\Delta G = \pi r^2 t \Delta G_v + 2\pi r t \gamma_E \tag{6.1}$$

where γ_E is the surface energy at the edge of the disc. The maximum value of ΔG occurs at a critical radius r_c found as in the three-dimensional case considered in section 5.3 by differentiating eqn. (6.1) to obtain the maximum value of ΔG. This gives

$$r_c = \gamma_E / \Delta G_v \tag{6.2}$$

Clusters of atoms less than r_c in radius are unstable.

Using the same approximation for the temperature dependence of the free energy change as on p. 105, namely

$$\Delta G_v = \Delta H_v \left(\frac{T_E - T}{T_E} \right) \tag{6.3}$$

where ΔH_v is the latent heat per unit volume, gives

$$r_c = \frac{\gamma_E}{\Delta H_v} \frac{T_E}{(T_E - T)} \tag{6.4}$$

Taking typical values for the quantities in eqn. (6.4) shows that the critical two-dimensional nucleus contains in the order of 100 atoms for a 5°K undercooling. The occurrence of such groups in solutions or vapours is unlikely until considerable supersaturation is attained. The fact that growth does occur at low supersaturation can be explained on the basis that atoms become attached at steps on the crystal surface where screw dislocations emerge, thus avoiding the need for two-dimensional nucleation.

In contrast the packing of the atoms in a liquid is only slightly less dense than in the solid and groups of critical size should occur frequently on the surface of crystals growing from the melt.

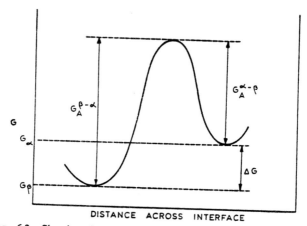

Fig. 6.2. Showing the relationship between the activation energy barrier at an interface and the net free energy change, accompanying the movement of an atom across the interface. $G_A^{\beta-\alpha}$ and $G_A^{\alpha-\beta}$ are the activation free energies for an atom to move from β to α and vice-versa respectively.

Hence, there should be no shortage of steps. Similarly in solids there should be no shortage of dislocations and surface discontinuities due to mutual interaction of neighbouring grains and it is reasonable to regard diffusional growth as controlled by the rate at which single atoms cross the interface. In the following treatment of the growth from the solid or liquid, two-dimensional nucleation and the other difficult problems about the nature of the interface and mechanism of atom attachment are ignored and growth is assumed to be a single atom process. The growth

THEORY OF DIFFUSIONAL GROWTH PROCESSES 157

of coherent phases is thus excluded because they require the coupled movement of groups of atoms. The energy relations for the thermally activated transfer of single atoms across the interface at a temperature T below T_E are shown in Fig. 6.2. $G_A^{\alpha-\beta}$ and $G_A^{\beta-\alpha}$ are the free energy of activation for an atom to cross the interface from α to β and from β to α respectively.

The number of atoms leaving α per unit area of interface per unit time is

$$p.A_\beta.N_{s(\alpha)}.v_\alpha.e^{-G_A^{\alpha-\beta}/kT}$$

where $N_{s(\alpha)}$ is the number of atoms per unit area of α at the interface, v_α is the vibration frequency in α, $p_\alpha(\sim \frac{1}{6})$ the probability that a vibration is in the correct direction, and A is the accommodation coefficient for the β crystal, i.e. the fraction of the number of sites on the surface at which atoms can be accommodated into the growing lattice.

Similarly the number of atoms leaving β per unit area is

$$p_\beta.A_\alpha.N_{s(\beta)}.v_\beta.e^{-G_A^{\beta-\alpha}/kT}$$

where the symbols have the same meaning as before, but referred to the other phase. The net rate of accumulation of atoms on the surface of the β crystal is

$$p_\alpha.A_\beta.N_{s(\alpha)}.v_\alpha.e^{-G_A^{\alpha-\beta}/kT} - p_\beta A_\alpha.N_{s(\beta)}.v_\beta e^{-G_A^{\beta-\alpha}/kT}$$

The rate of advance of the interface $\dot{R}$ is this expression times the volume of one atom in β, λ. To simplify the algebra it is convenient to use the following approximations:

$$N_{s(\alpha)} \simeq N_{s(\beta)} = N_s$$
$$v_\alpha \simeq v_\beta = v$$
$$p_\alpha \simeq p_\beta = p$$

Then

$$\dot{R} = \lambda.p.N_s.v(A_\beta e^{-G_A^{\alpha-\beta}/kT} - A_\alpha e^{-G_A^{\beta-\alpha}/kT}) \quad (6.5)$$

At the equilibrium temperature T_E, the free energies of α and β are equal, and so $G_A^{\alpha-\beta} = G_A^{\beta-\alpha}$ and $\dot{R} = 0$. Raising the temperature above T_E makes $\dot{R}$ negative, i.e. the β crystal disappears; at temperature below T_E β grows at a finite rate.

Since the difference between the activation free energies for the forward and backward reaction is equal to the free energy of reaction ΔG as shown in Fig. 6.2, it is possible to put eqn. (6.5) into the form

$$\dot{R} = \lambda.N_s.v.p[A_\beta e^{-G_A^{\alpha-\beta}/kT} - A_\alpha e^{-(G_A^{\alpha-\beta} + \Delta G)/kT}] \quad (6.6)$$

In the case of polymorphic changes in solids it is reasonable to assume that the accommodation coefficients are the same on both sides of the interface, i.e.

$$A_\alpha = A_\beta = A \quad (6.7)$$

and in this case eqn. (6.6) simplifies to

$$\dot{R} = \lambda.N_s.v.p.Ae^{-G_A^{\alpha-\beta}/kT}(1 - e^{-\Delta G/kT}) \quad (6.8)$$

For small undercooling $\Delta G \ll kT$ and the exponential in the brackets may be expanded as a series

$$\dot{R} = \lambda.N_s.v.p.Ae^{-G_A^{\alpha-\beta}/kT}\left(1 - 1 + \frac{\Delta G}{kT} - \tfrac{1}{2}\left(\frac{\Delta G}{kT}\right)^2 + \ldots\right)$$

Neglecting squared and higher terms gives

$$\dot{R} = \lambda.N_s.v.p.Ae^{-G_A^{\alpha-\beta}/kT}\frac{\Delta G}{kT} \quad (6.9)$$

and using eqn. (6.3) for ΔG gives

$$\dot{R} = \lambda.N_s.v.p.Ae^{-G_A^{\alpha-\beta}/kT} \cdot \frac{\Delta H.\Delta T}{T_E.kT} \quad (6.10a)$$

or

$$\dot{R} = \lambda.N_s.v.p.Ae^{-G_A^{\alpha-\beta}/kT} \cdot \frac{\Delta S}{kT}.\Delta T \quad (6.10b)$$

where ΔT is the undercooling, ΔH the latent heat of transformation and ΔS the entropy change.

Further, for solid state changes it is plausible to take the activation energy for movement of an atom across the boundary,

THEORY OF DIFFUSIONAL GROWTH PROCESSES 159

$G_A^{\alpha-\beta}$ as that for self-diffusion along a high angle α grain boundary. $v.p.e^{-G_A^{\alpha-\beta}/kT}$ is then the grain boundary diffusion coefficient. $D_{G.B.(\alpha)}$ giving

$$\dot{R} = \lambda . N_s A . D_{G.B.(\alpha)} . \left(\frac{\Delta G}{kT}\right) . \Delta T \qquad (6.10c)$$

The eqns. (6.10) are expressions for the linear rate of growth of a crystal growing from another crystalline phase at small undercoolings. For the growth from a melt the approximation of equal accommodation coefficient is inadmissible and eqn. (6.6) must be used. The only simplification possible is to put $G_A^{\alpha-\beta}$ as that for diffusion in the melt.

It follows from eqns. (6.6) and (6.10) that the rate of growth is zero at T_E (when $\Delta G = 0$) and also at 0°K, attaining a maximum at some intermediate temperature. This behaviour is similar to that found for the rate of nucleation. The physical interpretation is much the same. At temperatures approaching T_E the rate is slow because the driving forces is negligible; at low temperature although the driving force is high the forces restraining atomic movement are also high. The temperature of maximum growth represents the optimum balance between driving force and restraint.

Jackson and Chalmers† calculated $\dot{R}$ as a function of T for a copper crystal growing from the melt using eqn. (6.6) and found a maximum value of 250 cm/sec at an undercooling of approximately 300°C. In view of the magnitude of the growth rate and also of the fact that homogeneous nucleation occurs at a temperature 100°C higher than that of maximum growth rate (see p. 106) it is evidently physically impossible to preserve liquid copper to room temperature by quenching.

The description of the growth of crystals from a liquid given above is a considerable over-simplification of this complicated and as yet imperfectly understood process. Several difficulties concerning the nature of the interface and the mechanism of

---

† K. A. Jackson and B. Chalmers, *Canad. J. Phys.* **34**, 473 (1956).

atom attachment have been glossed over. Moreover, eqn. (6.10) giving growth rate proportional to driving force is true only for this simple model.

In the derivation of the growth rate equation it was assumed implicitly that the temperature is uniform throughout both phases. In fact the latent heat of transformation is released at the interface warming it to a slightly higher temperature than the rest of the system. The actual interface temperature is determined by a steady-state condition in which the rate of heat evolution equals the rate of conduction away from the interface. The latent heat of solidification is high ($\sim$ 2 kcal/mole) and so the effect is important during the growth of crystals from the melt. In fact it is clear that at undercoolings of a few degrees or less the rate of growth is completely determined by the rate of heat extraction. In the case of solid state phase changes the latent heat is smaller by a factor of ten and the effect may be neglected.

6.3. Growth of a Single Phase in Two Component Systems

The diffusional growth of a phase in a two component system involves not only the transfer of atoms across an interface but also the re-distribution of the species since the growing phase must have a different composition from the parent. The growth of precipitates from supersaturated solid solutions (continuous precipitation) is a common example. The rate of growth depends upon the rate at which atoms are brought to, or removed from, the interface by diffusion and the rate at which they cross the interface. During the early stage of growth the interface reaction must be the slower of these steps because of the limited area of interface and because the distance over which diffusion is necessary tends to zero. At large particle sizes the reverse is true because the diffusional flux gets progressively slower due to the continuous removal of solute from solution reducing the concentration gradient which is the driving force for diffusion, whilst the flux across the interface increases due to the increase in area. It is convenient to consider first the two limiting cases; growth limited only by the interface reaction and then growth limited only by diffusion.

Growth controlled by the interface process

In this model it is assumed that diffusion in the solution is very much faster than the interface process so that the composition of the solution remains uniform throughout. Consider the growth of a precipitate with solute concentration c_β. Let the concentration of solute in the solution in equilibrium with the precipitate, neglecting size effects, be c_E. Provided the supersaturation in the solution is small the increase in the partial molar free energy of the solute in the supersaturated solution over that at concentration c_E is negligible in comparison with the energy barrier at the interface and it is permissible to assume that the activation energy for crossing the interface is independent of the concentration in the solution. In this case the net gain of atoms per unit area by the precipitate ψ is proportional to the difference between the concentration existing at the interface and the equilibrium value† c_E

$$\psi(t) = \psi_0[c(t) - c_E] \quad (6.11)$$

where ψ_0 is a constant and $c(t)$ the concentration in α at time t.

The linear rate of growth is

$$\dot{R} = \psi/(c_\beta - c_E) \quad (6.12)$$

The concentration in solution decreases as precipitation progresses and can be expressed in terms of y, the fraction of the available solute actually precipitated at time t.

$$1 - y = \frac{c(t) - c_E}{c_I - c_E} \quad (6.13)$$

where c_I is the initial concentration in the solution. Therefore

$$\dot{R}(t) = \frac{\psi_0}{(c_\beta - c_E)}[(c_I - c_E)(1 - y)] \quad (6.14)$$

---

† In more supersaturated solutions ψ is a function of $(c(t) - c_E)$ and of the variation of partial molar free energy of solute with composition.

Thus the rate of growth decreases continuously during precipitation. The complete solution of eqn. (6.14) in terms of $\dot{R}$ as a function of t requires a knowledge of $y(t)$ which can only be derived in terms of the overall kinetics of nucleation and growth. This will be considered in the next chapter. At small times when $(1 - y) \simeq 1$ the rate of growth is constant and a graph of x against t is a straight line. At later stages the growth rate decreases and the graph falls below the initial linear portion.

Determination of $\dot{R}_y$ at fixed fractional precipitation y, enables the temperature variation of ψ_0 to be studied. The slope of a Arrhenius graph of log $\dot{R}_y$ vs. $1/T$ yields the activation energy for the interface reaction, provided that c_E does not vary significantly with temperature.

Growth controlled by diffusion

In this model it is assumed that the rate of removal of atoms from the solution at the interface is very much faster than the rate at which atoms arrive at the interface. Thus the concentration in the solution at the interface is maintained at the equilibrium value c_E. Again c_E is regarded as independent of precipitate size.

Consider an isolated spherical particle of precipitate of radius R and solute concentration c_β growing in an infinitely large, homogeneous supersaturated solid solution of initial composition c_I. The concentration conditions are shown schematically in Fig. 6.3. As the interface advances a small distance dR the number of atoms captured by the precipitate per unit area of interface is $c_\beta . dR$. Of these $c_E . dR$ were present at the interface, the remainder being supplied by diffusion. From Fick's law the diffusive flux at the interface is $D.(\partial c/\partial r)_{r=R}$ where D is the value of the diffusivity of solute in the matrix at the concentration c_E and $(\partial c/\partial r)_{r=R}$ is the concentration gradient at the interface. Therefore the rate of growth, dR/dt, is given by

$$(c_\beta - c_{(E)})\left(\frac{dR}{dt}\right) = D\left(\frac{\partial c}{\partial r}\right)_{r=R} \qquad (6.15)$$

THEORY OF DIFFUSIONAL GROWTH PROCESSES 163

($\partial c/\partial r$) is found from the appropriate solution of Fick's equation using a time dependent value of R. A simpler approximate solution is possible when the supersaturation is small. In this case depletion of solute extends to distances very large compared to the radius of the particle and the rate of increase of R is very small. It is then reasonable to adopt a steady state approximation to the solution of the diffusion equation, physically equivalent to assuming that the concentration distribution around the growing

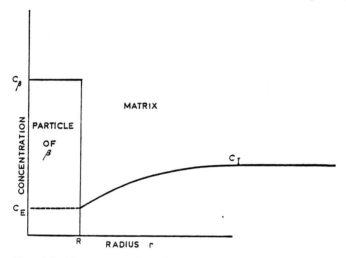

FIG. 6.3. The concentration of solute in and around a growing precipitate.

particle is the same as that which would exist if the particle absorbed atoms without increasing in size. This assumption of a quasi-steady state is frequently used in diffusion problems.

The steady state solution of Fick's equation for diffusion through a spherical shell of inner radius R and outer radius R_2, with D independent of c, is

$$\frac{\partial c}{\partial r} = \frac{c_2 - c_E}{(1/R - 1/R_2)} \cdot \frac{1}{r^2} \qquad (6.16)$$

where c_2 is the concentration at R_2.

Taking R_2 to be infinity and $c_2 = c(t)$ gives for the gradient at the interface

$$\left(\frac{\partial c}{\partial r}\right)_{r=R} = \left(\frac{c(t) - c_E}{R}\right) \qquad (6.17)$$

Combining eqns. (6.15) and (6.17) gives

$$(c_\beta - c_E)(dR/dt) = D(c(t) - c_E)/R \qquad (6.18)$$

i.e.

$$R\left(\frac{dR}{dt}\right) = D\left(\frac{c(t) - c_E}{c_\beta - c_E}\right) \qquad (6.19)$$

In general, precipitation occurs on a number of particles each of which draws solute from the solution. This competition leads to a continuous decrease in the matrix concentration from c_I and thus in the rate of growth of each particle. In the case of small supersaturations the number of particles is small and so the distance of separation is large compared to the ultimate size. In the analysis of this problem presented by Zener and Wert† it was assumed that under these conditions the concentration remote from each particle is approximately equal to the average concentration $c(t)$ throughout the solution at time t. Further $c(t)$ is related to the fraction y already precipitated at time t by the equation

$$\frac{c(t) - c_E}{c_I - c_E} = (1 - y) \qquad (6.20)$$

The rate of growth for one of a set of competing particles is then given by

$$R\left(\frac{dR}{dt}\right) = D\left(\frac{c_I - c_E}{c_\beta - c_E}\right)(1 - y) \qquad (6.21)$$

Equation (6.21) can only be integrated when y is known as a function of time. At small times $(1 - y) \simeq 1$ and R^2 is proportioned to the time and the volume to $t^{3/2}$. A graph of R against

---

† C. Zener and C. Wert, *J. Appl. Phys.* **21**, 5 (1950).

t is parabolic initially but eventually falls away as competition becomes important.

This approximate and somewhat arbitrary approach was later fully justified by Ham† in a more rigorous analysis of the diffusion problem. He showed that in a solid solution of initially uniform composition of low supersaturation where the ultimate precipitate particle size is small compared with their separation, the solute concentration in the solution is virtually identical to the steady-state field for a particle of fixed radius and that the concentration decreases uniformly except within small volumes around the particles. Since these volumes may be neglected the ratio of the instantaneous supersaturation $(c(t) - c_E)$ to the initial value $(c_I - c_E)$ is equal to the fraction of solute remaining in super-saturated solution $(1 - y)$. This is the assumption used in writing eqn. (6.20).

The situation in heavily supersaturated solutions is less clear. According to Zener‡ solute depletion is then confined to a comparatively thin shell of radius R_2 surrounding the growing particles and the concentration gradient may be regarded as approximately linear, as shown in Fig. 6.4. The expression for the rate of growth is derived as follows. As the radius R increases by dR the radius of the depleted sphere R_2 increases by dR_2. Mass balance requires the equality of the shaded areas in Fig. 6.4. Thus

giving
$$\tfrac{4}{3}\pi R^3 (c_\beta - c_I) = \tfrac{4}{3}\pi (R_2^3 - R^3)\tfrac{1}{2}(c_I - c_E) \quad (6.22)$$

$$R_2^3 = R^3 \, 2\frac{(c_\beta - \tfrac{1}{2}c_E - \tfrac{1}{2}c_I)}{(c_I - c_E)} \quad (6.23)$$

The concentration gradient $(\Delta c/\Delta r)$ is

$$\frac{\Delta c}{\Delta r} = \frac{c_I - c_E}{R_2 - R} = \frac{c_I - c_E}{\dfrac{R 2^{1/3} (c_\beta - \tfrac{1}{2}c_E - \tfrac{1}{2}c_I)^{1/3}}{(c_I - c_E)^{1/3}} - R} \quad (6.24)$$

---
† F. S. Ham, *J. Chem. Phys. of Solids*, **6**, 335 (1958).
‡ C. Zener, *J. Appl. Phys.* **20**, 962 (1949).

Putting this into eqn. (6.15) gives

$$R\frac{dR}{dt} = D\frac{(c_I - c_E)}{\left[\dfrac{2^{1/3}(c_\beta - \tfrac{1}{2}c_E - \tfrac{1}{2}c_I)^{1/3}}{(c_I - c_E)^{1/3}} - 1\right]}\left(\frac{1}{c_\beta - c_E}\right) \quad (6.25)$$

Integrating gives

$$R^2 = 2D\left[\frac{c_I - c_E}{c_\beta - c_E}\right]\left[\dfrac{1}{\dfrac{2^{1/3}(c_\beta - \tfrac{1}{2}c_E - \tfrac{1}{2}c_I)^{1/3}}{(c_I - c_E)^{1/3}} - 1}\right] \quad (6.26)$$

Again a parabolic relation is predicted for the rate of growth of an isolated particle. For an array of competing particles, there

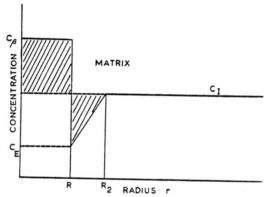

FIG. 6.4. Zener's approximation to the solute distribution around a precipitate growing from a highly supersaturated solution.

is no fundamentally justified means of allowing for the mutual interference of neighbouring particles as in the case of dilute solutions. If the concentration distribution is as in Fig. 6.4 competition takes the form of impingement of the depleted shells. The usual method of allowing for this effect is to multiply either eqn. (6.25) or eqn. (6.26) by the arbitrary factor $(1 - y)$.† These are not equivalent but give similar results up to approximately $y = \tfrac{1}{2}$ diverging thereafter. Other authors‡ have suggested that the

---
† C. Wert, *J. Appl. Phys.* **20**, 943 (1949).
‡ T. Mishima, *Proc. 1st World Met. Congress*, 668, Am. Soc. Metals, 1951.

impingement factor should be $(1 - y)^2$ particularly when precipitation is heavily localized so that competition is more effective than in a random distribution.

So far attention has been devoted to spherical particles. It was for long thought that the time dependence of non-spherical particles would be quite different. For example, Zener and Wert[†] suggested that whilst the radius of a cylindrical precipitate is proportional to $t^{1/2}$ the length should be proportional to t, and thus the volume to t^2. The basis of the idea was that lengthwise growth takes the particle into regions of the solution not affected by the radial diffusion so that the concentration conditions at the tip and thus the rate of growth remains constant. Similarly it was proposed that the radius of a disc is proportional to t and the thickness to $t^{1/2}$ giving the volume proportional to $t^{5/2}$. Ham,[‡] however, showed that the concentration conditions at the edge of cylinders and discs do not remain constant during diffusion controlled growth. By approximating cylinders and discs to prolate and oblate spheroids respectively, he was able to show that provided the shape remains constant during growth all dimensions are proportional to $t^{1/2}$, and the volume to $t^{3/2}$. When the rate of growth varies with direction due to some surface process, the shape of the particle, as defined by the eccentricity of the spheroid changes. For a disc of constant thickness growing radially only, the radius is related to time and the volume to t^2. The length and volume of a cylinder of constant radius is proportional to t.

6.4. Growth Dependent upon both Diffusion and the Interface Process

In the more general case when diffusion and the interface reaction are of comparable rates, the concentration of solute in the solution at the interface is maintained at a value between c_E

[†] C. Wert, *J. Appl. Phys.* **20**, 943 (1949); C. Zener, *ibid.* **20**, 962 (1949).
[‡] F. S. Ham, *op. cit.*; *J. Appl. Phys.* **30**, 1518 (1959).

and the average value in the solution $c(t)$, the actual value representing a balance between the rate at which atoms arrive at the interface and the rate at which they are removed.

Consider one isolated particle, which for convenience is assumed to be spherical. However the discussion applies equally to all shapes provided that growth is three-dimensional, the shape is invariant and the same kinetic law applies to all dimensions. The flux across the interface into the precipitate is

$$\psi(t) = \psi_0(c_B - c_E) \qquad (6.27)$$

where c_B is the concentration maintained at the interface and the flux to the interface is

$$J = D\left(\frac{\partial c}{\partial r}\right)_{r=R} \qquad (6.28)$$

Employing the steady state approximation to Fick's law eqn. (6.17)

$$\left(\frac{\partial c}{\partial r}\right)_{r=R} = (c(t) - c_B)(1/R) \qquad (6.29)$$

Equating the right-hand sides of eqns. (6.27) and (6.28) and using (6.29) gives

$$\psi_0(c_B - c_E) = D(c(t) - c_B)(1/R)$$

Solving for c_B gives

$$c_B = \frac{Dc(t) + \psi_0 c_E R}{\psi_0 R + D} \qquad (6.30)$$

When $D \gg \psi_0$ the reaction is interface controlled and

$$c_B = c(t)$$

When $D \ll \psi_0$, $\qquad c_B = c_E$.

The rate of growth is given from eqn. (6.18) with c_E replaced by c_B as

$$(c_\beta - c_E)(dR/dt) = (1/R)D(c(t) - c_B) \qquad (6.31)$$

Substituting for c_B gives

$$\frac{dR}{dt} = \frac{\psi_0 D(c(t) - c_E)}{(\psi_0 R + D)(c_\beta - c_E)} \qquad (6.32)$$

THEORY OF DIFFUSIONAL GROWTH PROCESSES 169

The transition to non-isolated particles is made as before by using Ham's result that the concentration remote from the precipitated $c(t)$, is continuously reduced uniformly in accordance with eqn. (6.20).

Therefore for an array of particles

$$\frac{dR}{dt} = \frac{\psi_0 D(c_I - c_E)}{(\psi_0 R + D)(c_\beta - c_E)}(1 - y) \qquad (6.33)$$

which can be integrated only when y is known as function of time.

During the early stages of reaction when y is small, depletion of the matrix is negligible and $(1 - y)$ is negligible. It follows that when R is small the term in R^2 is much smaller than that in R and so the radius is approximately proportional to t, i.e. the growth is interface controlled. At large sizes the squared term dominates and the growth law is characteristic of diffusion limited growth. This is in accord with the general idea discussed in section 6.3.

6.5. Coarsening of Precipitates

When a system containing a dispersed phase is annealed at a high temperature the number of particles of the dispersed phase decreases whilst the average particle size increases. Observation shows that during the coarsening process the larger particles grow at the expense of the smaller ones. The driving force for the change is the tendency to reduce the overall free energy of the system by reducing the total area of internal interfaces. The process requires diffusion of solute from regions close to small particles to regions around large ones, implying that the concentration of solute in the solution in equilibrium with a precipitate is larger for a small particle of precipitate than for a large one. This can be readily understood in a qualitative sense in terms of a free energy diagram shown in Fig. 6.5. Since the proportion of atoms located at the interface increases as size decreases it follows that the average free energy per atom of precipitate is greater for small particles. Of the two curves in Fig. 6.5 for the β phase the higher one refers to small particles and the lower one to large ones. Constructing the common tangents to the matrix curve shows that the solubility of small particles $C_{\alpha(s)}$ is larger than that of large particles $C_{\alpha(L)}$. In a system containing mixed sizes, concentration gradients exist in the matrix which promotes diffusion of solute from regions around small particles to those adjacent to large particles onto which it is precipitated. The removal of solute from the solution near to small particles causes the latter to dissolve.

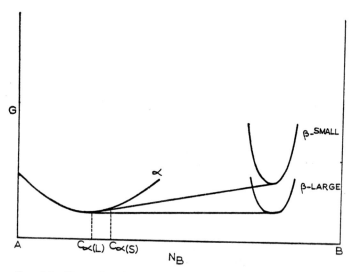

FIG. 6.5. Illustrating that the concentration of solute in a solution α increases as the radius of curvature of the precipitate particles decreases.

The dependence of solubility on size can be derived as follows. Consider two particles of radius r_1 and r_2, r_1 being smaller than r_2, containing N_1 and N_2 moles of one component B of a binary system. If the activity of B in the matrix in equilibrium with these particles is a_1 and a_2 respectively, the free energy change in transferring dN moles of B from the smaller to the larger particles is $dN(RT \ln a_2/a_1)$; this is equal to the change in the surface free energy $\gamma(dS_2 - dS_1)$ where dS_1 and dS_2 are the changes in the surface areas of the particles and γ is the interfacial free energy, assumed independent of the radius.

Now
$$dS_1 = 8\pi r_1 dr_1 \quad \text{and} \quad dS_2 = 8\pi r_2 dr_2$$
If the volume per mole in the precipitated phase is V then
$$V dN = 4\pi r_1^2 dr_1 = 4\pi r_2^2 dr_2$$
so that
$$dS_1 = 2(V/r_1)dN \quad \text{and} \quad dS_2 = 2(V/r_2)dN$$
Hence
$$RT \ln\left(\frac{a_2}{a_1}\right) = 2V\gamma \left(\frac{1}{r_2} - \frac{1}{r_1}\right) \tag{6.34}$$

In ideal solutions or in dilute solutions conforming to Henry's law a_2/a_1 is equal to c_2/c_1, the ratio of the solubilities of the two particles. Hence
$$RT \ln(c_2/c_1) = 2V\gamma \left(\frac{1}{r_2} - \frac{1}{r_1}\right) \tag{6.35}$$
Equations (6.34) and (6.35) are forms of the Thomson–Freundlich equation for the solubility as a function of particle size.

An analysis of the kinetics of growth during coarsening presents immense difficulties due to the complex composition distribution throughout the solution. Greenwood[†] succeeded in deriving the rate of growth for a very idealized model. His result is
$$R\left(\frac{dR}{dt}\right) = \frac{2Dc_\alpha V\gamma}{RT . c_\beta}\left(\frac{1}{R_M} - \frac{1}{R}\right) \tag{6.36}$$
where D is the diffusion coefficient of the solute and c_α is the concentration of solute in solution in equilibrium with a planar interface, c_β is the concentration in the precipitate and R_M is the arithmetic mean radius of the system of particles. It follows that particles smaller than the mean are dissolving at a rate which increases as size decreases; particles bigger than R_M are growing. A particularly interesting point is that dR/dt has a maximum value at $R = 2R_M$.

[†] G. W. Greenwood, *Acta Met.* 4 243 (1956).

The greater solubility of small precipitates is an example of a more general principle that the solubility of a metastable phase is greater than one of greater stability. In Fig. 6.6 the curve marked β^1 is the free energy curve for a metastable phase β^1. Since the curve must be higher than that for stable β the tangent to the α curve touches at a higher concentration of solute as shown. Once nuclei of the stable phase are formed the β^1 dissolves to supply the solute for growth of β in much the same way as

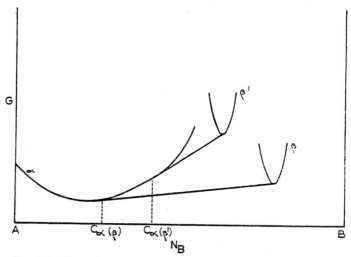

FIG. 6.6. Illustrating that the solubility of a metastable phase, β^1, is greater than that of a stable phase β.

discussed above. Many practically important examples of the behaviour are known including the formation of graphite in steels and white cast irons, in which cementite the metastable phase is more soluble than graphite.

6.6. Stress-assisted Precipitation on Dislocations

In the growth processes considered earlier the diffusion of solute to the growing phase was associated with differences in solute concentration in the matrix. However, the real driving force for migration is a difference in the partial molar free energy $\bar{G}$ of the species concerned between two points in the system. Concentration differences are but one way of inducing differences in $\bar{G}$. There are others. For example, a temperature gradient in a chemically homogeneous solution induces a gradient of the partial molar free energy of the solute and diffusion occurs along this temperature gradient even though the composition is uniform. The accommodation of solute atoms in solid solutions generally involves elastic distortion of the solvent lattice, either compressive or tensile depending upon the size of the solute atoms relative to the space into which it is inserted. In these circumstances an elastic strain gradient induces diffusion because the energy of a solute atom is reduced when it moves to a position in the lattice at which the strain is opposite to that associated with the atom. Dislocations give rise to strain gradients. In the case of a positive edge dislocation the lattice above the slip plane is in a state of compression and that below in tension. Consequently the energy of a solute atom associated with compressive strain in the lattice is reduced when it is located in the tensile region of a dislocation strain field. Conversely the energy of solutes producing tensile strain in the lattice is less for an atom in the compressive region of a dislocation than for one in a normal lattice site. Thus the partial molar free energy of solute atoms is less at dislocations than in regions remote from the defects and as a result solute atoms drift towards the dislocations forming "atmospheres" around the dislocations. The equilibrium distribution of solute in an unsaturated solution is microscopically non-uniform although its partial molar free energy is everywhere constant. In supersaturated solutions, precipitate particles nucleate from the atmospheres and further solute drains towards the dislocations, under the action of the stress field to feed the growing particles. The formation of atmospheres and precipitates on dislocations is an important part of theories of the yield point and strain and quench ageing in iron. The analysis that follows is restricted to a pure edge dislocation for simplicity although in practice it is probable that screw dislocations are more important in these phenomena.

The kinetics of drift to dislocation was first discussed by Cottrell and Bilby.† The interaction energy between a positive edge dislocation and a solute with polar coordinates r, θ, θ being measured from the slip direction is

$$U = A \frac{\sin \theta}{r} \qquad (6.37)$$

where A is a constant that depends upon the elastic constants of the material, the misfit between the atoms and the strength of the dislocation. The formula breaks down at the core (radius r_0) of the dislocation where the material does not behave as a Hookean solid. Large solute atoms are most strongly bound at $\theta = 3\pi/2$ and $r = r_0 \simeq 2$Å, i.e. immediately below the extra half plane of the dislocation. For carbon and nitrogen in α-Fe experiment indicates that $U \simeq 0.5$ eV a value in accord with that predicted by eqn. (6.37).

The force attracting a solute atom to a dislocation is the gradient of the interaction energy, dU/dr, just as in a mechanical system the force acting on a body is equal to the potential energy gradient. The mobility M of an atom is the velocity it attains under unit force. In Chapter 3, p. 87, it was shown that in ideal solutions $D = MkT$ where D is the diffusivity. Hence the velocity of an atom towards the dislocation is

$$v = M \frac{dU}{dr} = \frac{A}{r^2} \cdot \frac{D}{kT} \qquad (6.38)$$

in which the angular dependence of U has been neglected.

Atoms originally at a distance r from the dislocation require a time t to reach it given by

$$t = \frac{r}{v} = r^3 \frac{kT}{A \cdot D} \qquad (6.39)$$

When the time t has elapsed all solute atoms originally within a radius r have segregated. Considering unit length of dislocation

---

† A. H. Cottrell and B. A. Bilby, *Proc. Phys. Soc.* A62, 49 (1949).

the number of atoms within a distance r is $\pi r^2 \cdot l \cdot c_I$ where c_I is the concentration of solute in a uniform solution in terms of number of atoms per unit volume. Thus the number of atoms to have reached a unit length of dislocation, $n(t)$ at time t is

$$n(t) = c_I \pi \left(\frac{AD}{kT} \cdot t\right)^{2/3} \tag{6.40}$$

The derivation of eqn. (6.40) is highly simplified. The proper treatment given by Cottrell and Bilby gave $3(\pi/2)^{1/3}$ as the constant in place of π. The error, which in fact is negligible, arises from the assumption implicit in the above derivation that solute atoms follow a radial path to the dislocation, whereas a proper analysis shows that they trace out circular paths.

If the dislocation density is L lines/cm^2 (i.e. the total length of dislocations per unit volume of crystal), the number of atoms segregated per unit volume is $n(t).L$ and the fraction y of the original solute on dislocations is

$$y = \pi L \left(\frac{AD}{kT} \cdot t\right)^{2/3} \tag{6.41a}$$

and

$$\frac{dy}{dt} = \pi L \left(\frac{AD}{kT}\right)^{2/3} \cdot \frac{2}{3} \cdot t^{-1/3} \tag{6.41b}$$

This expression is valid for the early stages of ageing. Harper† modified the equation to allow for the fall in the concentration in the matrix as ageing proceeds (i.e. competition) by assuming that dy/dt is proportional to the fraction of solute left in the matrix, $(1 - y)$.

Thus

$$\frac{dy}{dt} = \pi L \left(\frac{AD}{kT}\right)^{2/3} \cdot \frac{2}{3} \cdot t^{-1/3}(1 - y) \tag{6.42}$$

Separating the variables and integrating gives

$$y = 1 - \exp -\left[\pi L \left(\frac{AD}{kT} \cdot t\right)^{2/3}\right] \tag{6.43a}$$

---

† S. Harper, *Phys. Rev.* **83**, 209 (1951).

Harper tested the equation for the segregation of carbon in α–Fe. He determined the fraction of solute still in solution using an internal friction technique and showed that log $(1 - y)$ is proportional to $t^{2/3}$ as eqn. (6.43a) predicts. The gradient of a graph of log $(1 - y)$ vs. $t^{2/3}$ is equal to $\pi L(AD/kT)$; evaluation of this slope for a series of temperatures enabled the activation energy for the process to be determined from an Arrhenius graph. The activation energy of 20,000 cal/mole is close to that for carbon diffusion in ferrite as the model demands.

The theory of Cottrell and Bilby neglects the fact that as soon as an appreciable fraction of solute is clustered around the dislocation back-diffusion occurs under the action of the concentration gradient, and further that the interaction energy progressively decreases as the dislocation strain is relaxed. Thus it might be expected that eqn. (6.43a) should hold only for initial stages of atmosphere formation and not at all for precipitation since the latter must involve very many atoms. However, in fact, the Harper modification to the Cottrell–Bilby equation is frequently in good agreement with the rates of precipitation. This suggests that the process of precipitation comprises first atmosphere formation followed by the nucleation of precipitate at intervals along the dislocation line, growth occurring by the drift of atoms into the atmosphere followed by very rapid diffusion down the core and capture by the precipitate. On this model drift in the stress field of the dislocation is the slowest step and capture into the precipitates effectively stops back-diffusion allowing eqn. (6.43a) to be obeyed.

Ham[†] and Bullough and Newman[‡] have made more detailed analyses taking into account back-diffusion and matrix depletion but not the reduction in the interaction energy. Ham assumed that the removal of atoms from the core into the precipitate is infinitely fast. The latter authors examined various rates of transfer across the interface. Both results differ significantly from Harper's except at small time when all models predict a $t^{2/3}$ dependence.

[†] F. S. Ham, *J. Appl. Phys.* **30**, 915 (1959).
[‡] R. Bullough and R. C. Newman, *Proc. Roy. Soc.* **A249**, 427 (1959); **A266**, 198 (1962).

6.7. Growth of Eutectoid and Cellular Precipitates

A eutectoid transformation is one in which a single phase decomposes during cooling to two new phases differing in structure and composition from each other and the parent phase. The product of such a transformation is often in the form of alternate lamellae of the two constituent phases. Pearlite in steels, an aggregate of ferrite and cementite, is the best known and technologically the most important example. A similar two-phase lamellar aggregate is produced by some precipitation reactions in non-ferrous systems. In these only one of the product phases is new, the other being structurally the same as the parent phase although differing in composition and orientation. These reactions are now referred to as cellular reactions although the term recrystallization reaction is sometimes used to emphasize that recrystallization of the parent phase must be involved to produce the new orientation. The two types of phase changes have much in common and may be considered together.

It is evident that eutectoid transformations are more drastic and complex than any considered before because they involve the simultaneous nucleation of two crystallographically different phases and the co-operative growth of two crystals of different composition behind a common incoherent interface with the parent phase. The mechanism is only imperfectly understood. Nucleation is almost always heterogeneous occurring virtually exclusively at grain boundaries of the parent phase or at inclusions. There has been considerable argument as to which constituent is the active nucleus. For example, in the case of pearlite it was long thought that cementite was the nucleus. However, it is now recognized that neither cementite nor ferrite alone is sufficient. Nucleation of a nodule of pearlite requires the establishment of stable nuclei of both; either may be the first to form chronologically depending upon the conditions, but it is not effective until the other is nucleated.†

---

† R. F. Mehl and W. C. Hagel, *Prog. in Metal Phys.* 6, Pergamon, (1956); J. W. Cahn and W. C. Hagel, *Decomposition of Austerite by Diffusional Processes*, Wiley, 1962.

178 THE KINETICS OF PHASE TRANSFORMATIONS IN METALS

Growth of individual nodules proceed both along the grain boundaries (until impingement with neighbouring nodules) and radially into the contiguous parent grains. Until recently, the generally accepted model of growth involved both longitudinal growth in the direction of the axes of the lamellae (edgewise growth) and lateral growth by repeated nucleation of the two phase alternatively (sideways growth). However, careful metallographic work, particularly by Hillert,† has shown that for pearlite at least, sideways nucleation is negligible and that each nodule is essentially a bi-crystal being composed of two interwoven crystals one of ferrite and one of cementite. Hence, in considering the rate of growth only edgewise growth need be considered.

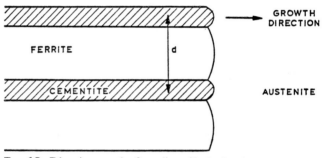

FIG. 6.7. Edgewise growth of pearlite. d is the interlamellar spacing.

Edgewise growth of pearlite is illustrated in Fig. 6.7. d, the average repeat distance of the plates in the direction normal to the ferrite and carbide plates, is termed the interlamellar spacing of the nodule. Observation shows that a growing nodule quickly establishes a constant spacing, characteristic of the temperature of the reaction; and also that d decreases as the temperature is decreased below the equilibrium transition temperature. Experi-

---

† M. Hillert, *Decomposition of Austenite by Diffusional Processes*, Wiley, 1963.

mentally it is found that there is an approximate linear relation between d and $1/(T_E - T)$, T_E being the equilibrium transition temperature. Growth requires the diffusion of carbon atoms from the ferrite interface to the cementite interface. This may occur either within the austenite, along the interface, within the product phase or by a combination of all three processes. Whatever the path, the diffusion conditions remain constant once a constant spacing is established. That is to say, the distance over which diffusion is necessary, related to the spacing, and the composition parameters do not change. Hence the rate of growth is constant for a constant temperature. This is borne out by experiment. However, it is observed that nodules are often sharply asymmetrical, from which it follows that the rate of growth is faster in some directions than in others. In particular, a nodule often grows much faster into one of the grains with which it is in contact than into the other, suggesting that the crystallographic relationship between austenite and ferrite affects the rate of propagation of the interface.

It follows that the rate of growth and the interlamellar spacing are interdependent. Some independent condition is needed to relate the two. Zener† pointed out that part of the free energy released by transformation is absorbed as the energy of the interfaces between the ferrite and cementite plates. When the spacing is small the interfacial energy accounts for a large proportion of the free energy change and little is left to provide a driving force. On the other hand, at large spacings the distance through which diffusion of carbon is necessary to accomplish the partition is unduly large and leads to a slow rate of growth. The system, therefore, finds a compromise intermediate spacing. Zener suggested that the spacing realized is the one that leads to a maximum rate of growth.

The free energy associated with the ferrite–cementite interface per unit volume of pearlite is

$$\Delta G_s = \frac{2\gamma}{d} \tag{6.43b}$$

---

† C. Zener, *Trans. A.I.M.E.* **167**, 550 (1956).

THE KINETICS OF PHASE TRANSFORMATIONS IN METALS

where γ is the interfacial energy per unit area. The net free energy change accompanying the formation of unit volume of pearlite is

$$\Delta G = \Delta G_0 - \frac{2\gamma}{d} \tag{6.44}$$

ΔG_0 is the free change in free energy in the absence of interfaces. The minimum possible spacing d_M is that for which $\Delta G = 0$, i.e.

$$d_M = \frac{2\gamma}{\Delta G_0} \tag{6.45}$$

Putting eqn. (6.45) into (6.44) gives for ΔG

$$\Delta G = \Delta G_0(1 - d_M/d) \tag{6.46}$$

Assuming that the rate of growth, $\dot{R}$, is determined only by the rate at which diffusion accomplishes the necessary redistribution of carbon then $\dot{R}$ must be proportional to the diffusion coefficient of carbon and inversely to the spacing d because d defines the distance over which diffusion is necessary. If it is further assumed that the rate of growth is proportional to the driving force, ΔG, so that

$$\dot{R} = K\Delta G \frac{D}{d} \tag{6.47}$$

K is a proportionality constant and $\dot{R}$ is the rate of interface movement in one particular direction in the growing nodule. Substituting for ΔG from eqn. (6.46) gives

$$\dot{R} = KD \frac{\Delta G_0}{d} \left(1 - \frac{d_M}{d}\right) \tag{6.48}$$

$\dot{R}$ has a maximum when

$$\frac{d\dot{R}}{dd} = 0$$

Thus

$$\frac{d\dot{R}}{dd} = KD \frac{\Delta G_0}{d^2} \left(2 \frac{d_M}{d} - 1\right) = 0 \tag{6.49}$$

i.e.

$$d = 2d_M \tag{6.50}$$

The actual spacing is twice the minimum value. Using eqn. (6.45) gives

$$d = \frac{4\gamma}{\Delta G_0} \tag{6.51}$$

The free energy change ΔG_0 is approximately linearly proportional to the undercooling. It may be written (see p. 105)

$$\Delta G_0 = \frac{\Delta H(T_E - T)}{T_E} \tag{6.52}$$

Therefore

$$d = \frac{4\gamma \cdot T_E}{\Delta H(T_E - T)} \tag{6.53}$$

where ΔH is the latent heat of transformation per unit volume of pearlite.

Thus a graph of log d against log $(T_E - T)$ should be linear of slope -1. The experimental values often conform to this prediction. The numerical values calculated from eqn. (6.53) using the known values of ΔH, γ and T_E agree within a factor of 3 to 10 with the measured values. Although this agreement is fair more detailed considerations suggest that the Zener condition for d is an over simplification. Other conditions have been suggested.†

Having obtained values for d the next step is to derive the rate of growth. To be rigorous this requires solution of the diffusion equation for the particular system. However, a simple approximate method will be given here. Assuming that the partition of carbon occurs by diffusion within the austenite the current of carbon atoms to the tip of the cementite plate J is approximately given by

$$J = \frac{D^\gamma}{\alpha d}(c_f^\gamma - c_c^\gamma) \tag{6.54}$$

---

† J. W. Cahn and W. C. Hagel, *op. cit.*

where c_c^γ and c_f^γ are concentrations in the austenite just ahead of the cementite and ferrite respectively and $\alpha d (\alpha \sim \frac{1}{2})$ is the diffusion distance. c_c^γ and c_f^γ are the compositions on the extrapolated phase boundaries in the phase diagram as shown in Fig. 6.8.

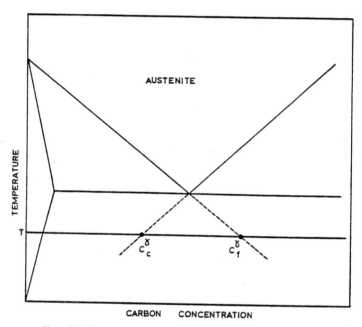

Fig. 6.8. Extrapolation of the austenite phase boundaries.

As the ferrite plate extends a distance dR the loss of carbon in front of the ferrite is $dR(c_c^\gamma - c_f^\alpha)$ where c_f^α is the carbon concentration in ferrite.

Therefore

$$\dot{R}(c_f^\gamma - c_f^\alpha) = \frac{D(c_f^\gamma - c_c^\gamma)}{\alpha d} \qquad (6.55)$$

The values of $\dot{R}$ calculated from this equation are in reasonable agreement with the measured rates of growth of pearlite in certain commercial steels but not in pure iron–carbon alloys, alloy steels or cellular reactions.

This model is not completely satisfactory because if carbon diffusion was the rate determining step then a nodule would grow equally fast in all directions. The observation of asymmetrical nodules implies that the interface is less mobile in certain orientations than in others suggesting that interface mobility is important. There is now plenty of evidence that the rate of the interface reaction is also important and that diffusion is predominantly down the austenite–pearlite interface as is to be expected since the diffusivity is much larger in this region. A more sophisticated analysis of the growth of eutectoids taking into account these effects has been given by Cahn.†

A simple alternative expression for $\dot{R}$ may also be derived as follows. ΔG and $1/d$ are both proportional to $(T_E - T)$. Thus, eqn. (6.47) may be written

$$\dot{R} = K_1 D(T_E - T)^2$$
$$= K_2 e^{-E_A/kT}(T_E - T)^2 \qquad (6.56)$$

where K_1 and K_2 are constants and E_A the activation energy for carbon diffusion. Equation (6.56) predicts that $\dot{R}$ has a maximum value at some critical undercooling, and this is borne out by experiment.

Further Reading for Chapter 6

Thermodynamics in Physical Metallurgy, American Society Metals, 1950.
Atom Movements, American Society Metals, 1951.
Phase Transformations in Solids, Wiley, 1951.
The Decomposition of Austenite by Diffusional Processes, Wiley, 1963.
TURNBULL, D., *Solid Static Physics*, 3 (Eds. F. Seitz and D. Turnbull), Academic Press (1956).
Liquid Metals and Solidification, *Am. Soc. Met.* (1958).
U. MARTIUS, *Prog. in Met. Phys.*, 5, 279 (1954).
W. C. WINEGARD, *Met. Rev.* 6, 57 (1961).

---

† J. W. Cahn, *Acta Met.* 7, 18 (1959).

CHAPTER 7

The Kinetics of Diffusional Transformations

7.1. Introduction

Formally the kinetics of a constant temperature transformation are described by an equation of the general type

$$y = f(t) \qquad (7.1)$$

where y is the fractional transformation at time t and $f(t)$ is some function of t. The Johnson–Mehl, the Austin–Ricketts and the autocatalytic equations discussed in sections 2.5 and 2.6 are particular forms of eqn. (7.1) which have been found empirically to describe the kinetics of many transformations in metals. In this chapter the problem of deriving $f(t)$ for certain types of diffusional transformation is discussed.

At any time t, dy/dt depends upon (a) the number of domains of product phase, which in turn is determined by the time dependence of the rate of nucleation I in the interval $t = 0$ to t, (b) the rate of growth of each domain which depends upon the geometry of the domains and the rate of growth of each dimension. In principle $f(t)$ can be deduced once explicit expressions are derived for these two factors. In fact it is possible to carry out this procedure rigorously for a few very simple models only because as noted in section 5.6 and 5.12 it is generally impossible to deduce the time dependence of I and so it is necessary to make arbitrary assumptions concerning the nucleation kinetics. Furthermore the effect of the mutual interference of the growing crystals can be allowed for explicitly only in idealized cases. In this chapter attention is confined to some models for which reasonably satisfactory analyses are possible.

7.2. The Diffusion Controlled Growth of a Fixed Number of Crystals from a Slightly Supersaturated Solid Solution

The transformation analysed most successfully is one in which a fixed number of nuclei of product domains are formed all at the same time, ($t = 0$), with no other nuclei appearing at any subsequent time, and in which the growth rate is determined by the rate of diffusion of solute through the matrix.† This model is likely to be a reasonable approximation to many continuous precipitation processes.

Diffusion controlled growth was discussed in section 6.3. The radial rate of growth of a spherical domain dR/dt was found to be

$$R \left(\frac{dR}{dt}\right) = D \left(\frac{c_I - c_E}{c_\beta - c_E}\right)(1 - y) \quad (7.2)$$

where the symbols have the same meaning as in section 6.3.

If N domains of product per unit volume nucleated at $t = 0$ with a random distribution, and the mean radius of all the domains at t is R, then the number of atoms removed from unit volume of the solution at t is $N\frac{4}{3}\pi R^3(c_\beta - c_E)$. The number available for precipitation is $(c_I - c_E)$ and so the fraction transformed is given by

$$y = N\tfrac{4}{3}\pi R^3 \left(\frac{c_\beta - c_E}{c_I - c_E}\right) \quad (7.3)$$

Hence

$$R = \left[\frac{3}{4\pi} \frac{1}{N} \frac{(c_I - c_E)}{(c_\beta - c_E)}\right]^{1/3} y^{1/3}$$

and

$$\frac{dR}{dt} = \frac{1}{3}\left[\frac{3}{4\pi} \frac{1}{N} \frac{(c_I - c_E)}{(c_\beta - c_E)}\right]^{1/3} y^{-2/3} \cdot \frac{dy}{dt}$$

Using these relations to eliminate R from eqn. (7.2) gives

$$\frac{dy}{dt} = 3D \left(\tfrac{4}{3}\pi N\right)^{2/3} \left[\frac{(c_I - c_E)}{(c_\beta - c_E)}\right]^{1/3} y^{1/3}(1 - y) \quad (7.4)$$

---

† C. Zener, *J. Appl. Phys.* **20**, 962 (1949); and C. Wert, *ibid.* **21**, 5 (1950); F. S. Ham, *J. Chem. Phys. of Solids*, **6**, 335 (1958).

Separating the variables and integrating using the boundary condition that $y = 0$ at $t = 0$ gives

$$\frac{1}{6}\left[\ln\frac{1 + y^{1/3} + y^{2/3}}{(1 - y^{1/3})^2}\right] + \frac{1}{\sqrt{3}}\left[\tan^{-1}-\left(\frac{2y^{1/3} + 1}{\sqrt{3}}\right) - \tan^{-1}-\frac{1}{\sqrt{3}}\right] = (\tfrac{4}{3}\pi N)^{2/3} D \left(\frac{c_I - c_E}{c_\beta - c_E}\right)^{1/3} t \quad (7.5)$$

Equation (7.5) gives a sigmoidal y—log t graph (Fig. 7.1).

The rate constant in eqn. (7.5) is

$$k = D(\tfrac{4}{3}\pi N)^{2/3} \left(\frac{c_I - c_E}{c_\beta - c_E}\right)^{1/3} \quad (7.6)$$

The temperature dependence of the reaction rate arises from the variation of k with temperature. Thus the empirical activation energy E_A is given by

$$E_A = \frac{d(\ln k)}{d(1/T)} = \frac{d(\ln D)}{d(1/T)} + \frac{2}{3}\frac{d(\ln N)}{d(1/T)} + \frac{1}{3}\frac{d}{d(1/T)}\left[\ln\frac{c_I - c_E}{c_\beta - c_E}\right] \quad (7.7)$$

The first term on the right-hand side of eqn. (7.7) is the activation energy for the diffusion process that governs the reaction rate. The second term depends upon the temperature dependence of the nucleation rate; it may be positive or negative depending upon whether the temperature is below or above the nose of the nucleation C-curve. The third term which depends upon the form of the equilibrium diagram, is likely to be negligible compared with the first term unless c_E or c_β varies sharply with temperature. It follows that in general E_A determined for the overall reaction cannot be directly related to the activation for the governing diffusion process. An illustration of this conclusion was given by Wert for the precipitation of cementite from supersaturated ferrite.† The value of E_A derived from the temperature dependence of the rate of precipitation in specimens quenched directly to the reaction temperatures in the range 25° to 315°C was

---

† C. Wert, *J. Appl. Phys.* **20**, 943 (1949).

13,000 cal/mole. In a further series of measurements specimens were first quenched to 27°C, held for a time sufficient to allow 5 per cent of the available carbon to be precipitated and then up-quenched to various temperatures. In this case $E_A = 17,500$ cal/mole. The reason for this difference is that in the first series

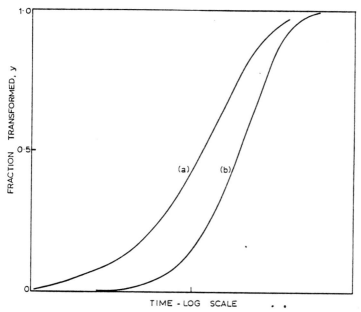

FIG. 7.1. Isothermal transformation curves according to eqns. (7.5) (diffusion controlled growth) curve (a) and (7.10) (interface controlled growth) curve (b).

nucleation was accomplished at various temperatures whereas in the second it took place at the same temperature in all samples and so N was constant and $d(\ln N)/d(1/T)$ zero. The activation energy for diffusion of carbon in ferrite is 18,400 cal/mole suggesting that, in this case, the contribution of the concentration term in eqn. (7.7) is negligible.

7.3. The Growth of a Fixed Number of Crystals from a Slightly Supersaturated Solution at a Rate Influenced by the Interface Reaction

It was shown in section 6.3 that when the capture process at the crystal-matrix interface is much slower than the diffusion of atoms to the interface, the radial rate of growth is given by

$$\frac{dR}{dt} = \psi_0 \left(\frac{c_I - c_E}{c_\beta - c_E}\right)(1 - y) \tag{7.8}$$

Combining with eqn. (7.3) gives for the rate of reaction

$$\frac{dy}{dt} = 3\psi_0 \left(\frac{4\pi N}{3}\right)^{1/3} \left(\frac{c_I - c_E}{c_\beta - c_E}\right)^{2/3} y^{2/3}(1 - y) \tag{7.9}$$

Integrating, noting that $y = 0$ at $t = 0$,

$$\frac{1}{6}\left[\ln \frac{1 + y^{1/3} + y^{2/3}}{(1 - y^{1/3})^2}\right] - \frac{1}{\sqrt{3}}\left[\tan^{-1}\frac{(2y^{1/3} + 1)}{\sqrt{3}} - \tan^{-1}\frac{1}{\sqrt{3}}\right]$$

$$= \psi_0(\tfrac{4}{3}\pi N)^{1/3}\left(\frac{c_I - c_E}{c_\beta - c_E}\right)^{2/3} t \tag{7.10}$$

eqn. (7.10) gives a sigmoidal rate curve of greater slope than the diffusion controlled model (see Fig. 7.1).

The integrated rate equation for a model in which diffusion and the interface process are comparable is derived in the same way as eqns. (7.5) and (7.10) starting from eqn. (6.32) for the rate of growth. The sigmoidal rate curve has no unique slope, depending upon the relative magnitudes of ψ_0 and D. The limiting slopes are those for eqns. (7.5) and (7.10) shown in Fig. 7.1.

7.4. The Diffusion Controlled Growth of a Variable Number of Crystals from Supersaturated Solid Solutions

A more general model than those considered previously is one in which nucleation proceeds concurrently with growth. The difficulty is to select a realistic time dependence for the nucleation. As noted in section 5.6 it is quite impossible to derive this for homogeneous nucleation. Classical theory predicts an increasing rate of nucleation during a short transient period. In addition a rapid decrease is expected for a continuous precipitation reaction due to the progressive reduction in the supersaturation. The assumption of heterogeneous nucleation does little to reduce the uncertainty (cf. section 5.12) because of the impossibility of expressing W as a function of y.

In view of these remarks the only approach possible is to derive the kinetics on the basis of arbitrary assumptions concerning the time dependence of I. Although it is comparatively simple to derive the rate equation for any assumed $I(t)$ it is impossible to integrate it in closed form and numerical methods are required. Only one case has been solved.† It is found that the effect of continuing nucleation is to increase the steepness of the sigmoidal rate curve by an amount which depends upon the form of $I(t)$. This is better illustrated in terms of the approximate treatment discussed in the next section.

7.5. Approximate Treatments of Precipitation Kinetics

The equations derived in previous sections of this chapter involve complicated mathematical expressions and, as noted in section 7.4, it is very difficult to extend the analysis. Fortunately it is possible to use approximate treatments that both facilitate comparison with experimental data and extension to other models.

Growth without nucleation. At small values of y the amount of solute drained from the solution is small and so the effect of competition is negligible. Thus the rate of growth, approximately that of an isolated particle, is obtained from eqn. (6.21) by placing $y = 0$, i.e.

$$R\left(\frac{dR}{dt}\right) \simeq D\left(\frac{c_I - c_E}{c_\beta - c_E}\right) \quad (7.11)$$

or

$$R^2 \simeq 2D\left(\frac{c_I - c_E}{c_\beta - c_E}\right)t \quad (7.12)$$

---

† J. Burke, *Phil. Mag.* **6**, 1439 (1961).

from which it follows that the time dependence of the rate of growth is

$$\frac{dR}{dt} \simeq \sqrt{(D/2)} \left(\frac{c_I - c_E}{c_\beta - c_E}\right)^{1/2} t^{-1/2} \quad (7.13)$$

Combining eqns. (7.11), (7.12) and (7.13) with (7.3) gives

$$\frac{dy}{dt} \simeq 4\sqrt{2\pi} ND^{3/2} \left(\frac{c_I - c_E}{c_\beta - c_E}\right)^{1/2} t^{1/2} \quad (7.14)$$

The effect of competition for solute during later stages is allowed for by putting an arbitrary factor $(1 - y)$ into eqn. (7.14).

$$\frac{dy}{dt} \simeq 4\sqrt{2\pi} ND^{3/2} \left(\frac{c_I - c_E}{c_\beta - c_E}\right)^{1/2} t^{1/2}(1 - y) \quad (7.15)$$

Integration of eqn. (7.15) gives

$$y \simeq 1 - \exp - \left[\tfrac{8}{3}\sqrt{2\pi} ND^{3/2}\left(\frac{c_I - c_E}{c_\beta - c_E}\right)^{1/2} t^{3/2}\right] \quad (7.16a)$$

which may be written

$$y = 1 - \exp[-(kt)^{3/2}] \quad (7.16b)$$

Equation (7.16) which is a Johnson–Mehl equation with time exponent $n = \tfrac{3}{2}$ is compared with the more accurate eqn. (7.5) in Fig. 7.2 by plotting log log $(1/1 - y)$ against log t. Equation (7.16) is linear of slope 3/2. Equation (7.5) is a curve but it is noted that up to about $y = 0.4$ the deviation from linearity is slight and so in this range of y eqn. (7.16) is a good approximation to eqn. (7.5).

It is to be noted that the value of n in eqn. (7.16) is the same as the exponent of t in the expression for the volume of a growing particle when $y \simeq 0$. In the example above $R \propto t^{1/2}$ and vol $\propto t^{3/2}$. When the rate limiting step is the interface reaction then the rate of growth of an isolated crystal is constant (eqn. 6.12). Thus $R \propto t$, vol $\propto t^3$ and $n = 3$. The particular form of Johnson–Mehl equation for this model is obtained by replacing eqn. (7.11) by eqn. (6.12) and following the same steps that led to eqn. (7.16).

THE KINETICS OF DIFFUSIONAL TRANSFORMATIONS 191

Consequently it is relatively simple to evaluate n in an equation corresponding to eqn. (7.16b) for various models using the growth equations given in Chapter 6—see Table 7.1. The value of $n = 3/2$ is characteristic of the diffusion controlled growth of a fixed number of crystals irrespective of the shape of the crystals,

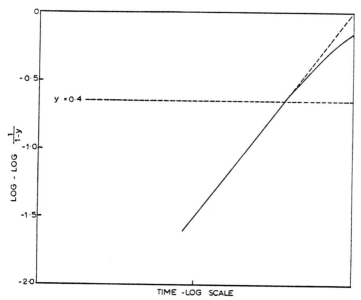

FIG. 7.2. Graph of log log $(1/(1 - y))$ against log t for eqn. (7.5)—full line—and eqn. (7.16)—broken line.

provided that the shape does not change during growth (cf. section 6.3). $n = \tfrac{2}{3}$ is associated with stress assisted precipitation [eqn. (6.40)]. Values of n between $\tfrac{3}{2}$ and 3 occur when diffusion and the interface process are of comparable rate and neither is completely controlling.

Nucleation and growth. As an example consider the case in which the rate of nucleation per unit volume is a constant I, and the rate of growth is diffusion controlled.

At time t, the radius of a crystal nucleated at time t_1 ($0 < t_1 < t$) is

$$R^2 \simeq 2D \left(\frac{c_I - c_E}{c_\beta - c_E}\right)(t - t_1) \qquad (7.17)$$

If v is the volume of this particle, the rate of volume growth at t is

$$\frac{dv}{dt} \simeq 4\pi R^2 \cdot \frac{dR}{dt} = 4\sqrt{2}\pi D^{3/2} \left(\frac{c_I - c_E}{c_\beta - c_E}\right)^{3/2} (t - t_1)^{1/2} \qquad (7.18)$$

TABLE 7.1.

Some values of n, the exponent of t, in the approximate rate equation, $y \simeq 1 - \exp - (kt)^n$.

Model	n
Diffusion controlled growth of a fixed number of particles†	3/2
Growth of a fixed number of particles limited by the interface process†	3
Diffusion controlled growth of cylinders in axial direction only†	1
Diffusion controlled growth of discs of constant thickness†	2
Growth on dislocations‡	2/3
Nucleation at a constant rate and diffusion controlled growth	5/2
Growth of a fixed number of eutectoid cells	3
Nucleation at a constant rate and growth of a eutectoid	4

In most precipitation reactions the volume occupied by the precipitate is small compared with that of the matrix and so may be neglected. Thus, the number of nuclei formed from t_1 to $t_1 + dt_1$ is $I dt_1$ and the rate of increase of volume of this group of crystals at t is $I dt_1 \cdot dv/dt$. The increase of volume V of all particles formed from $t = 0$ to $t = t$ is

---

† F. S. Ham, *J. Chem. Phys. of Solids*, **6**, 335 (1958); *J. Appl. Phys.* **30**, 1518 (1959).

‡ A. H. Cottrell and B. A. Bilby, *Proc. Phys. Soc.* **A62**, 49 (1949); F. S. Ham, *J. Appl. Phys.* **30**, 915 (1959).

$$\frac{dV}{dt} \simeq 4\sqrt{2\pi}D^{3/2}\left(\frac{c_I - c_E}{c_\beta - c_E}\right)^{3/2} I \int_0^t (t - t_1)^{1/2} dt_1$$

$$\simeq \frac{8\sqrt{2}}{3}\pi D^{3/2}\left(\frac{c_I - c_E}{c_\beta - c_E}\right)^{3/2} I.t^{3/2} \quad (7.19)$$

To convert volume of precipitate to fractional precipitation, eqn. (7.19) must be multiplied by $(c_\beta - c_E)/(c_I - c_E)$. The competition factor $(1 - y)$ is also placed in eqn. (7.19). Thus

$$\frac{dy}{dt} \simeq \frac{8\sqrt{2}}{3}\pi D^{3/2}\left(\frac{c_I - c_E}{c_\beta - c_E}\right)^{1/2} I.t^{3/2}(1 - y)$$

which integrates to

$$y \simeq 1 - \exp - \left[\frac{16}{15}\sqrt{2\pi}D^{3/2}\left(\frac{c_I - c_E}{c_\beta - c_E}\right)^{1/2} I.t^{5/2}\right] \quad (7.20)$$

This equation is a Johnson–Mehl equation with an exponent of $n = \frac{5}{2}$. It is a useful approximation to the accurate solution for values of $y < 0.4$. It is to be noted that the effects of continuing nucleation is to increase the time exponent over the value obtaining for growth of a fixed number of crystals. In general if the nucleation rate is given by

$$I = \text{const } t^b \quad (7.21)$$

where b is a constant then the time exponent n contains a contribution of $(b + 1)$ associated with nucleation.

7.6. Eutectoid Transformations and Transformations in Pure Solid Metals

It was shown in sections 6.2 and 6.7 that the linear rate of growth of a crystal growing during a polymorphic transition in pure metals and also of eutectoid domains is constant. This case differs from the ones considered in earlier sections of this chapter in two ways. (*a*) Interference between growing domains takes the form of direct impingement. Thus the growth rate of a domain is constant up to the time at which it impinges on another domain when it becomes zero. (b) The volume of the growing phase is not negligible in comparison with that of the matrix.

Consider unit volume of parent and let y_v be the volume of product formed at time t (i.e. y_v is the volume fractional transformation.) Let the rate of growth, which is the rate at which the product-parent interface moves normal to itself, be denoted by G and for the sake of simplicity suppose the growing domains are spheres.

The overall rate of increase of volume of product is

$$\mathrm{d}y_v/\mathrm{d}t = G.A \qquad (7.22)$$

where A is the total area of product–parent interface that is growing freely in the sample.

In order to evaluate A as a function of y_v and t, Johnson and Mehl† and Avrami‡ introduced the idea of an extended volume V_x or surface A_x defined as the volume or surface area of product that would have existed if all domains had grown without impingement and domains had continued to nucleate everywhere in the sample including the transformed volume. In other words V_x is the volume of the actual domains plus that of the "phantom" domains that would have nucleated in the transformed regions if they had been untransformed.

V_x and A_x are related by an equation similar to eqn. (7.22)

$$\mathrm{d}V_x/\mathrm{d}t = GA_x \qquad (7.23)$$

For a random distribution of nuclei the fraction of A_x that is not within a domain and hence is "free" interface area, is equal to the untransformed volume fraction, $1 - y_v$, i.e.

$$A = (1 - y_v)A_x \qquad (7.24)$$

Combining eqns. (7.23) (7.24) and (7.22) gives

$$\mathrm{d}y_v/\mathrm{d}V_v = 1 - y_v \qquad (7.25)$$

---

† W. A. Johnson and R. F. Mehl, *Trans. A.I.M.E.* **135**, 416 (1939).
‡ M. Avrami, *J. Chem. Phys.* **7**, 1103 (1939).

Separating the variables and integrating
$$-\ln(1 - y_v) = V_x$$
i.e.
$$y_v = 1 - e^{-V_x} \tag{7.26}$$

The volume of domains nucleated at time t_1 is
$$G^3(t - t_1)^3$$

If I is the rate of nucleation per unit volume of parent then the number of domains formed between t_1 and t_1 and $t_1 + dt_1$ is
$$dn = I(1 - y_v)dt_1 \tag{7.27}$$

The number of phantom nuclei that would have formed in the transformed volume is $Iy_v dt_1$ and thus the total number including phantoms is
$$dn^1 = I dt_1 \tag{7.28}$$

Hence the extended volume at t is
$$V_x = G^3 \int_0^t (t - t_1)^3 . I . dt_1 \tag{7.29}$$

Combination of eqns. (7.26) and (7.29) gives the equation for y_v as a function of t. For example, if N nuclei are formed at $t = 0$ and none thereafter, t_1 and dt_1 are zero and
$$V_x = (Gt)^3 N$$
and
$$y_v = 1 - e^{-NG^3 t^3} \tag{7.30}$$

which is a Johnson–Mehl equation with exponent $n = 3$.

If the rate of nucleation I is a constant
$$V_x = \tfrac{1}{4}IG^3 . t^4$$
and
$$y_v = 1 - e^{-\frac{1}{4}IG^3 t^4} \tag{7.31}$$

which is a Johnson–Mehl equation with exponent $n = 4$. This is the model originally treated by Johnson and Mehl.

The kinetics of grain boundary nucleated eutectoid reactions have been analyzed in detail by Cahn.† The results take the same form as equation (7.31) with various values of the time exponent.

---

† J. W. Cahn, *Acta. Met.* **4**, 449, (1956);—and W. C. Hagel, *Decomposition of Austenite by Diffusional Processes*, Wiley 1963.

CHAPTER 8

Martensitic Transformations

8.1. Definition of Martensitic Transformations

The name martensite was originally given to the very hard, acicular constituent produced by quenching steels from above the A_3 temperature. Investigation revealed that in medium and high carbon steels the transformation of face-centred cubic austenite to body-centred tetragonal martensite occurs at temperatures below about 200°C and that the time taken for an individual martensite crystal to grow to final size is less than 10^{-4} sec. This extraordinarily high growth rate could not possibly be attained at such low temperature if diffusion is responsible for the phase change, and thus it was suggested that each martensite crystal forms by shear of a volume of austenite containing many thousands of atoms. Subsequently it was found that a number of other solid state phase changes in both pure metals and alloys exhibit features similar to this transformation in steels. It is now usual to describe all such transformations as martensitic, and the product of them as martensite irrespective of structure.

The early attempts to distinguish between diffusional reactions (which used to be given the name nucleation and growth reactions) and martensitic changes were based upon kinetic considerations or sometimes upon the microstructural appearance or properties of the product. For example, it was suggested that all martensites must be very hard and that they can only form athermally, i.e. as the temperature is changing. However, this approach had to be abandoned when it was realized that several transformations which occurred at such temperatures and velocities that diffusion could not possibly be involved did not conform to these definitions of a martensitic change. Thus it became necessary to direct attention to the mechanism by which the structural change is accomplished. Consequently, a martensitic transformation is defined as one in which the growth of the product crystals takes place by the systematic coordinated movement of many atoms of the parent crystal, in such a way as to generate the product structure, the distance through which any one atom moves in the course of the transformation being a fraction of one lattice spacing. Any atom has the same neighbours in the product phase as in the parent crystal, only the relative positions are different. Thus the chemical composition of parent and product is identical and if the parent is ordered so also is the product. Furthermore, a martensite crystal contains the same number of atoms as the region of the parent crystal from which it is generated but, since there is a change of lattice, the shape and volume of the martensite crystal is different. If a surface of the parent crystal is polished flat before the transformation the shape change causes the martensite crystals to tilt out of the surface, as shown in Fig. 8.1. Observation of these surface reliefs is one of the main experimental criteria of a martensitic reaction.

The shape change is shown in idealized form in Fig. 8.1. *EFGH* is an initially plane surface of a parent crystal. *ABCDOLMN* is a martensite crystal bounded by, and linked with, the parent crystal by the interfaces *ABML* and *CDNO*. Due to the shape change the surface is tilted about the lines *AB* and *CD*. Measurement has shown that the *macroscopic or total shape change*, that is the change measured across an entire martensite plate, is equivalent to an invariant plane strain. An invariant plane strain is one in which each point in the body moves in the same direction through a distance proportional to the distance of the point from some fixed plane. In simple shear, the direction of movement is parallel to the fixed plane. The shape deformation in a martensite transformation may be regarded as a simple shear parallel to the plane of the plate plus a contraction or expansion normal to this plane. It follows that the surface of the plate must remain flat, although tilted relative to adjoining crystals.

198   THE KINETICS OF PHASE TRANSFORMATIONS IN METALS

An initially straight fiducial mark, SS^1, drawn on the surface prior to the transformation, remains continuous but is fragmented into three straight segments ST, TT^1 and T^1S^1.

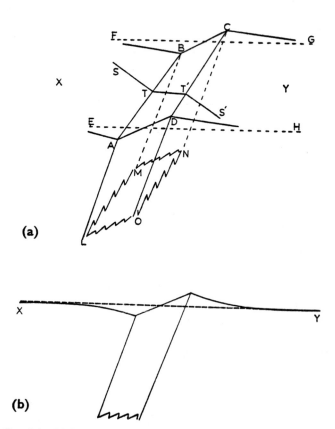

FIG. 8.1. (a) The shape change associated with a martensite plate. (b) Section through XY. (After Bilby and Christian.)

A section through XY shown in Fig. 8.1b illustrates that there is considerable elastic distortion induced in the parent around the plate due to the preservation of continuity at the interface. This distortion imposes constraints on the growing crystal and causes it to adopt the form of lenticular plates, the usual form of martensite crystals. Occasionally a single parent crystal is transformed by the passage of a single plane interface but in most cases several lenticular plates arise from a single grain giving the characteristic acicular microstructure.

Although the total shape deformation is equivalent to a homogeneous strain, the same strain when applied to the parent lattice does not generate the product lattice. It is necessary to invoke additional minor deformations, heterogeneous on an atomic scale, to account for the actual positions of the atoms in the martensite crystal. These may take the form of internal slip or twinning. Figure 8.2 illustrates how slip or twinning, which are *lattice invariant deformations*, may be combined with a homogeneous lattice deformation to produce various total shape deformations.

The plates formed in one parent crystal lie parallel to a certain plane of the parent lattice which is called the *habit plane*; it is usually a crystallographic plane of high indices. For example the habit plane of martensite in high carbon steels is $\{225\}_\gamma$. The habit plane often varies with reaction temperature and composition. Because of the co-ordinated nature of the atom movements there is always an orientation relationship between product and parent.

Martensitic reactions which occur on cooling are crystallographically reversible on heating. The plates within one parent grain revert to the parent phase to reproduce the single crystal from which they formed and on subsequent cooling the plates frequently form in exactly the same plane and with the same shape as in the first transformation. In some systems a side reaction intervenes during heating and prevents the reverse change. For example, quenched steels temper before the martensite can revert to austenite.

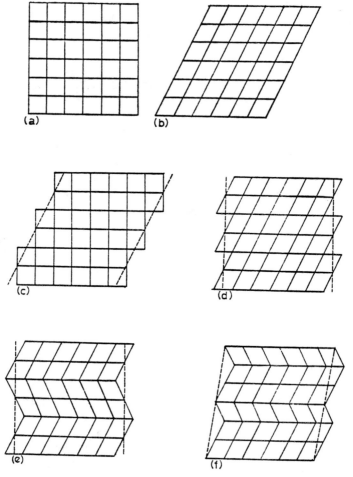

Fig. 8.2. The crystallographic features of a martensitic reaction: (a) the original lattice, (b) the lattice after a homogeneous shear, (c) slip plus zero lattice deformation giving the same shape change as in (b), (d) lattice deformation plus slip giving zero shape change, (e) lattice deformation plus uniform twinning giving zero shape change, (f) lattice deformation plus non-uniform twinning giving a finite shape change. (After Bilby and Christian.)

8.2. Thermodynamics of Martensite Transformations

No composition change accompanies the formation of martensite and it is permissible to regard the reaction as a phase change in a single component system. In this respect martensite formation is simpler to treat than diffusional changes in which re-distribution of the components occurs. The free energy–temperature relations are shown schematically in Fig. 8.3. γ is the parent phase stable

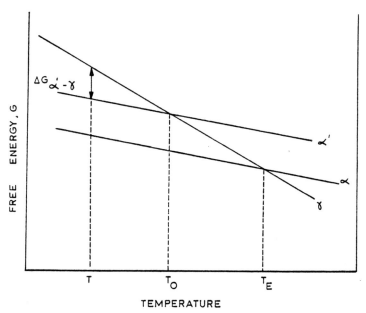

FIG. 8.3. Schematic representation of the free energy as a function of temperature for a high temperature phase γ, the phase stable at low temperature α and the product of a martensitic change α^1.

at high temperature and α is the stable phase at low temperature. The free energy–temperature curve for α intersects that for γ at T_E, the equilibrium transition temperature. α^1 is a phase of the same composition as γ formed by a diffusionless change. Its free energy is greater than that of the stable structure at all temperatures. At some temperature T_0 below T_E $G_{\alpha^1} = G_\gamma$. The formation of α^1 is thermodynamically impossible above T_0, but thermodynamically possible at any temperature below T_0 since only then is $\Delta G_{\alpha^1 - \gamma} = (G_{\alpha^1} - G_\gamma)$ negative. $\Delta G_{\alpha^1 - \gamma}$ is the driving force for the martensitic change to α^1.

The formation of a martensitic plate generates substantial elastic strain on account of the shape change and the constraints of the surrounding matrix. This strain energy absorbs part of the free energy released by the transformation and the net free energy change is only a fraction of ΔG. Further, the formation of a martensite nucleus involves the production of a new interface with associated energy. Surface and strain energies vary little with temperature whereas $|\Delta G_{\alpha^1 - \gamma}|$ increases as T decreases. Thus, further undercooling to a temperature M_s below T_0, is required to give sufficient driving force to initiate the change. This temperature, known as the martensite start temperature is the highest temperature at which a martensitic reaction can occur. $(T_E - M_s)$ varies from a few degrees centigrade in some pure metals to many hundred degrees in some complex alloy systems. An exact thermodynamic definition of M_s is not possible. It may be that temperature at which $|\Delta G_{\alpha^1 - \gamma}|$ becomes equal to the opposing forces, or it may be that temperature at which the activation energy for nucleation becomes sufficiently small.

Martensitic reactions may be induced at temperatures above M_s by an externally applied shear stress because the external stress tends to neutralize the internal stresses produced by the transformation. The higher the temperature above M_s the greater is the stress required to promote the transformation. There is a temperature M_d above which the change cannot be stress induced. M_d, which must be below T_0, is the temperature at which the stress required to induce transformation is just equal to the flow stress of the parent phase. At higher temperature the parent deforms plastically before the external stress reaches a level sufficient to induce transformation.

Considerable attention has been given to the problem of calculating T_0 and ΔG as a function of temperature because these are important quantities in any quantitative treatment of the kinetics. T_0 is related to the phase diagram as shown in Fig. 8.4

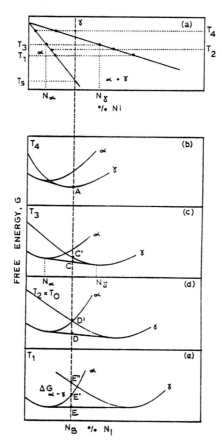

Fig. 8.4. (a) Part of the Fe–Ni phase diagram.
(b) (c) (d) and (e) are schematic free energy–composition diagrams for this system at four temperatures decreasing from top to bottom.
(After Kaufman and Cohen.)

for a system such as iron–nickel. The stable phase in this system is γ (f.c.c.) at high temperature and either α (b.c.c.) or a mixture of α and γ at low temperatures depending on the composition as shown in Fig. 8.4a. Figure 8.4,b c, d and e are schematic free energy composition curves for the two phases at four temperatures. At the highest temperature T_4 the stable state of an alloy, composition N_B, is γ, the free energy being A. At T_3 the stable state is a mixture of α, composition N_α, and γ, composition N_γ. N_α and N_γ are the points of contact of the common tangent to the two curves. Point C is the free energy of the mixture. Production of this state requires diffusion to accomplish the re-distribution of the iron and nickel atoms. If the rate of cooling to T_3 is sufficient to prevent diffusion then the actual free energy is C^1. At this temperature α of composition N_B cannot form because it has a higher free energy than C. Figure 8.4d is drawn for the temperature at which $G_\alpha = G_\gamma$ at the composition N_B. Again the lowest free energy D is derived from the rule of common tangents. If compositional changes during cooling to T_2 are prevented the actual free energy of the γ is D^1. This is also the free energy of α of the same composition and thus this temperature is T_0 for the alloy. In the bottom diagram the temperature is below T_0 and the change of γ to α of the same composition is accompanied by a decrease in free energy, ΔG. The final free energy E^1 is lower than that of $γ, E^{11}$, but still greater than that of the two-phase mixture E. If the alloy is quenched to T_1 it may transform martensitically to α if ΔG is greater than the opposing forces, i.e. if $T_1 < M_s$; or alternatively to $α + γ$ by a diffusional change. The transformation that actually occurs is the one that leads to the most rapid reduction in free energy. At low temperatures the rate of a diffusional change is virtually zero whereas a martensite change permits the rapid production of α.

At temperatures below T_s the equilibrium form is single phase α. At this, and all lower temperatures, a diffusionless reaction produces the thermodynamically stable phase without passing through the two-phase region. If $α^1$ is produced at a temperature at which it is metastable (e.g. T_1), it will become stable on subsequent cooling to T_s.

T_0 and ΔG may be evaluated provided the free energy–composition curves can be calculated theoretically or derived empirically from the phase diagram. Unfortunately there are insufficient data available for this approach to be used and an alternative, but fundamentally less satisfactory method is necessary. The free energy of γ, in an alloy A–B, of composition N_B is

$$G_\gamma = (1 - N_B)G_{A(\gamma)} + N_B G_{B(\gamma)}, + G_\gamma^M \quad (8.1)$$

where $G_{A(\gamma)}$ and $G_{B(\gamma)}$ are the free energies of the γ modification of pure A and pure B respectively and G_γ^M is the free energy of mixing of γ of composition N_B. A similar equation applies to α of the same composition:

$$G_\alpha = (1 - N_B)G_{A(\alpha)} + N_B G_{B(\alpha)} + G_\alpha^M \quad (8.2)$$

Subtracting eqn. (8.1) from (8.2) gives $\Delta G_{\alpha-\beta}$ the change in chemical free energy accompanying the diffusionless formation of α from γ

$$\begin{aligned}\Delta G_{\alpha-\gamma} &= (G_\alpha - G_\gamma) = (1 - N_B)[G_{A(\alpha)} - G_{A(\gamma)}] + N_B[G_{B(\alpha)} \\ &\quad - G_{B(\gamma)}] + (G_\alpha^M - G_\gamma^M) \\ &= (1 - N_B)\Delta G_{A(\alpha-\gamma)} + N_B \Delta G_{B(\alpha-\gamma)} + \Delta G_{\alpha-\gamma}^M \quad (8.3)\end{aligned}$$

where $\Delta G_{A(\alpha-\gamma)}$ and $\Delta G_{B(\alpha-\gamma)}$ are the differences in free energy between the α and γ allotropes of pure A and pure B respectively and $\Delta G_{\alpha-\gamma}^M$ is the difference between the free energies of mixing. $\Delta G_{A(\alpha-\gamma)}$ and $\Delta G_{B(\alpha-\gamma)}$ can be evaluated from specific heat measurements provided that the pure metals have α and γ forms. If they do not then the physical significance of the quantities is obscure and their values cannot be derived. In most cases the alloy is based upon one of the components as solvent and $\Delta G_{\alpha-\gamma}$ can be derived for at least this component. An example is the iron–nickel system. Iron is the solvent; γ-iron is f.c.c. and α-iron b.c.c. $\Delta G_{Fe(\alpha-\gamma)}$ has been calculated from specific heat data. Ni is f.c.c. at all temperatures and so $\Delta G_{Ni(\alpha-\gamma)}$ cannot be derived.

$\Delta G_{\alpha-\gamma}^M$ can be derived from activity measurements in α and γ. Where this data is not available it is necessary to assume models for the two solutions and use these as the basis of calculation. Such calculations have been made for the iron–nickel system as a function of nickel content. On the assumption that $\Delta G_{Ni(\alpha-\gamma)}$ is negligible it is found that $\Delta G_{\alpha-\gamma}$ at M_s is in the order of 300 cal/mole.

8.3. Kinetic Characteristics of Martensitic Reactions

A prerequisite for a martensite reaction is that the parent phase be retained unchanged to M_s. This requires some form of quenching to prevent diffusional decomposition at higher temperatures. Experimentally, it is found that the rate of cooling has to exceed some critical rate, the value of which depends upon the composition, grain size and thermal and mechanical history. The possible diffusional reactions have C-curve *TTT* diagrams and the initial cooling rate may be regarded as that necessary to bye-pass the nose of the fastest diffusional change. Once M_s is reached subsequent martensitic transformation may occur athermally or isothermally. In general, M_s is relatively unaffected by the cooling rate.

Athermal reactions. In most systems transformation occurs only during cooling and ceases if the temperature is held constant below M_s. The fraction transformed is a function of the temperature to which a specimen is cooled. The temperature at which complete transformation is realized is called M_f.

Growth of martensitic plates occurs extremely rapidly, the velocity being in the order of the speed of sound in the metal and virtually independent of the temperature. It follows that the activation energy for growth is effectively zero. Two limiting types of athermal reaction may be recognized. In the first, once a nucleus becomes active the interface advances rapidly until stopped by a grain boundary or some other structural obstacle in the parent crystal so that the plate attains final size almost instantaneously. The overall rate of transformation is determined by the number of nuclei and the size of the martensite plates and is independent of the rate of cooling. This behaviour occurs in systems in which there is a large barrier to the formation of nuclei of product. M_s is considerably below T_0 and so once a nucleus becomes active there is plenty of free energy available to overcome the forces opposing growth. It follows that these reactions exhibit extensive hysteresis between M_s for the cooling reaction and the start of the reverse change on heating called A_s. The martensitic reaction in an iron–30 per cent nickel alloy is an example of this type; the difference between M_s and A_s is of the order of 400°C.

The other limiting type involves the discontinuous growth of existing plates in addition to the nucleation of new ones as the temperature falls. The barrier to nucleation is small and M_s is close to T_0. Thus there is little free energy available to overcome the elastic strain. At any temperature a plate grows rapidly at roughly the same speed as in the previous case until the strain energy builds up to the point at which it equals the driving force when growth abruptly ceases. Further cooling makes available additional free energy enabling rapid growth to proceed until either a new balance point is reached or a structural obstacle blocks the interface. The overall rate of transformation is determined by the rate of cooling. An approximately equiatomic gold–cadmium alloy is an example; the difference between M_s and A_s is 16°C.

Isothermal martensitic reactions. Although the vast majority of martensitic reactions are athermal a number of isothermal examples are known. In some systems isothermal reaction follows athermal transformation when the cooling is stopped and the temperature held. Completely isothermal transformation occurs in a few alloys, and in these cases a *C*-curve *TTT* diagram results. The active temperature range is usually below room temperature as shown in Fig. 8.5 for an iron–nickel–manganese alloy. Isothermal martensite is structurally indistinguishable from athermal martensite. Growth is rapid, each plate reaching full size in a fraction of a second. Reaction proceeds by nucleation of new crystals rather than growth of existing ones. The reaction rate is governed by the rate at which nuclei appear and the size of the fully grown plates. The previous conclusion that growth is not thermally activated is not affected, but isothermal reactions show

208 THE KINETICS OF PHASE TRANSFORMATIONS IN METALS

that nucleation can be thermally activated in some systems.

The isothermal martensitic reaction observed in uranium-chromium alloys is rather exceptional in that it proceeds largely by slow growth of existing plates with virtually no new nucleation. In an isothermal reaction the driving force is constant and if growth is slow then the restraints must decrease with time. In this case it has been suggested that the elastic stresses are reduced by

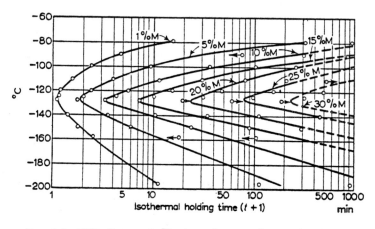

FIG. 8.5. TTT diagram for isothermal martensite reaction in an Fe–Ni–Mn alloy. (After Cech and Hollman.)

thermally activated plastic deformation in the product or matrix or both. Alternatively it is possible that the complexity of the atom movements renders the interface relatively immobile and some small scale readjustment of the atoms at the interface has to occur to enable growth to proceed. If this is so, then the reaction should be classed with the massive reactions mentioned briefly on p. 96.

Autocatalytic effects. The stresses generated around a martensite plate may induce the nucleation and growth of other plates nearby. This phenomenon known as the autocatalytic effect often produces "bursts" of transformation giving rise to discontinuities in the fraction transformed—temperature graph.

Thermal stabilization. If the cooling is interrupted above M_s, it is sometimes found that M_s is depressed. Similarly if the interruption occurs below M_s in an athermal reaction, transformation does not recommence immediately when the cooling is resumed. This retardation is referred to as thermal stabilization. It may affect isothermal as well as athermal reactions. Although it was thought that it constituted an essential characteristic of martensitic changes there is now good evidence that it only occurs in alloys containing interstitial solutes. A likely explanation is that during the holding the interstitial atoms, being comparatively mobile, migrate to form Cottrell atmospheres making the matrix more resistant to the atomic displacements involved in martensite formation.

8.4. The Nucleation of Martensite

From the previous discussion it is clear that an understanding of the kinetics of martensitic reactions depends upon a quantitative understanding of the rate of nucleation. Although several attempts have been made to compare calculated and measured nucleation rates no satisfactory theory is yet available. In the paragraphs following an indication is given of the approaches used.

The classical theory of homogeneous nucleation given in Chapter 5 may be applied directly. Suppose the embryos are in the form of oblate spheroids of radius r and thickness $2c$ (Fig. 8.6). If $r \gg c$ the volume is approximately $\frac{4}{3}\pi r^2 c$ and the surface area $2\pi r^2$. The free energy of formation of one embryo ΔG is

$$\Delta G = \tfrac{4}{3}\pi r^2 c \Delta G_v + 2\pi r^2 \gamma + \tfrac{4}{3}\pi r^2 c(A.c/r) \qquad (8.4)$$

where ΔG_v is the change in free energy per unit volume of martensite, i.e. $\Delta G_{\alpha-\gamma}$ in eqn. (8.3) expressed per unit volume of martensite. γ is the interfacial energy and A is an elastic strain energy constant defined so that Ac/r is the strain energy per unit volume of embryo. W the activation energy for the formation of a critical nucleus is that value of ΔG for which $\partial(\Delta G)\partial c = \partial(\Delta G)\partial r = 0$.

Differentiating eqn. (8.4) and applying these conditions gives

$$r_c = \frac{4A\gamma}{(\Delta G_v)^2} \quad (8.5)$$

$$c_c = -\frac{2\gamma}{\Delta G_v} \quad (8.6)$$

$$W = \frac{32}{3}\pi \frac{A^2\gamma^3}{\Delta G_v^4} \quad (8.7)$$

where r_c and c_c refer to the critical nucleus.

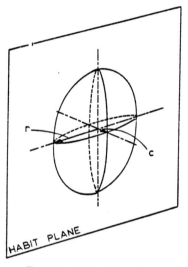

FIG. 8.6. Martensite embryo.

Since the activation energy for growth is negligible it is reasonable to assume that the equilibrium distribution of embryos is maintained even down to the lowest temperatures. Further, by analogy with eqn. (5.10) the rate of nucleation is given by

$$I = N.ve^{-W/kT} \quad (8.8)$$

because $U_I = O.v$ is in the order of the vibrational frequency. Equation (8.8) predicts a C-curve temperature dependence of I.

On this theory all martensitic reactions are essentially isothermal in nature. To account for athermal characteristics it is necessary to identify M_s as that temperature at which the isothermal nucleation rate becomes appreciable, say 1 nucleus/sec cm^3. ΔG_v increases with undercooling and from eqns. (8.7) and (8.8) it follows that I increases extremely rapidly as the temperature is decreased to the nose of the C-curve. At any temperature above

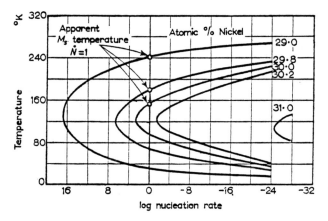

FIG. 8.7. Calculated isothermal nucleation rate as a function of temperature for Fe–Ni alloys. (After Fisher.)

M_s $I \ll 1$ and isothermal nucleation is not observed. Similarly below M_s $I \gg 1$ and isothermal nucleation cannot be suppressed by the most rapid quench. Fisher† applied this theory to the isothermal martensitic reaction in an iron–manganese–nickel alloy, the *TTT* curve of which is shown in Fig. 8.5, with remarkably close agreement between calculated and measured values.

The success of the approach was short lived. The isothermal nucleation rates calculated from eqn. (8.8) for various iron–nickel

---

† J. C. Fisher, *Trans. A.I.M.E.* **197**, 918 (1953); *Acta. Met.* **1**, 32 (1953).

alloys are shown in Fig. 8.7. In this theory the value of M_s for any alloy is that temperature for which log $I = 0$. It follows that any alloy for which the nucleation frequency at the nose of the C-curve is less than unity should not transform athermally during cooling to 0°K. Thus the theory predicts that alloys containing more than 30·1% Ni should not transform. In fact alloys containing up to 33% Ni transform athermally. Furthermore, Cohen† calculated r_c, c_c and W for an iron–nickel alloy containing 30% Ni in which M_s is − 40°C and $\Delta G_v = -315$ cal/mole or − 44 cal/cm³. A was calculated to be 500 cal/cm³ and a dislocation model of the austenite–martensite interface gave $\gamma = 200$ erg/cm². It was found that $r_c = 490$Å, $c_c = 22$Å and $W = 13 \times 10^7$ cal/mole of nuclei. Such an embryo would contain 2×10^6 atoms and the nucleation barrier at M_s would be 3×10^5 times larger than kT. The chances of such a nucleus arising by thermal activation is negligible.

The assumption of homogeneous nucleation is not an essential part of the classical theory. In fact experiment shows that nucleation is always heterogeneous. This approach has the advantage that it is unnecessary to suppose that thermal activation produces the critical nucleus. A number of sites exist in the parent having a spectrum of energy barriers. M_s is the temperature at which the driving force is sufficient to overcome the energy barrier of the most effective site. As the temperature is lowered the sites presenting larger barriers are rendered active by the greater driving force. Unfortunately when numerical comparisons are made as in the previous paragraph the same kind of disparity exists.

The strain-nucleus model is a special case of heterogeneous nucleation. The preferred sites are regarded as regions of matrix where local concentrations of stress exist. These " strain centres " may be regarded as groupings of dislocations enclosing a volume in which the lattice is intermediate between the parent and product. Growth occurs by the outward movement of the circumscribing dislocations. The theory in its original form

---

† M. Cohen, *Trans. A.I.M.E.* **212**, 171 (1958).

implies a gradual transition from one lattice to another whereas recent ideas suggest that the interface between the lattices is discontinuous. If this is so the lattice within the dislocations must be the product lattice and the frozen in embryos would have to be larger than the critical size of the classical theory in order to grow.

This theory predicts that an untransformed parent crystal contains regions of martensite up to about 500 Å radius. These would be detectable by electron microscopy. In fact no embryos have been detected.

Further Reading for Chapter 8

KAUFMAN, L. and COHEN, M., Thermodynamics and Kinetics of Martensitic Transformations, *Progress in Metal Physics*, Vol. 7, Pergamon Press, 1958.

BILBY, B. A. and CHRISTIAN, J. W., *The Mechanism of Phase Transformation in Metals and Alloys*, Inst. of Metals, 1956.

COHEN, M., The Nucleation of Solid-State Transformations, *Trans. A.I.M.E.* 171 (1958).

BILBY, B. A. and CHRISTIAN, J. W., The Crystallography of Martensitic Transformations, *J. Iron and Steel Inst.* **197**, 122 (1961).

OWEN, W. S. and GILBERT, A., The Kinetics of the Martensitic Transformation in Iron and Alloys, *J. Iron and Steel Inst.* **197**, 122 (1961).

COHEN, M., *Phase Transformations in Solids*, Wiley, 1951.

CHRISTIAN, J. W., *Decomposition of Austenite by Diffusional Processes*, Wiley, 1963.

Appendix A

Partition Functions

The denominator in the Boltzmann distribution law is called the Partition Function Q.

$$Q = \Sigma_i \, g_i e^{-\epsilon_i/kT} \qquad (A.1)$$

where the symbols are as defined in Chapter 1. The relationship between Q and some important thermodynamic functions are derived as follows.

Entropy S

Equation (1.26) is

$$S = k(\beta U_T + N \ln \Sigma_i e^{-\beta \epsilon_i}) \qquad (A.2)$$

Using the definition of β, eqn. (1.29) and of Q, eqn. (A.1) gives

$$S = U/T + kN \ln Q \qquad (A.3)$$

If the system contains one mole $N = \mathbf{N}$ (Avogadro's number), and the entropy per mole is

$$\underline{S = U/T + R \ln Q} \qquad (A.4)$$

where U is molar energy.

Helmholtz Free Energy F

Since $F = U - TS$
eqn. (A.3) rearranged gives

$$F = -NkT \ln Q \qquad (A.5)$$

or

$$F = -RT \ln Q \text{ per mole} \qquad (A.6)$$

Internal Energy U

The Gibbs–Helmholtz equation of chemical thermodynamics is

$$\frac{U}{T^2} = -\frac{\partial}{\partial T}\left(\frac{F}{T}\right)_V \tag{A.7}$$

Using eqn. (A.5) gives

$$U = NkT^2 \frac{\partial}{\partial T} \ln Q_v \tag{A.8}$$

Equilibrium Constant K

The equilibrium constant for the reaction

$$aA = bB \tag{A.9}$$

is

$$K = \frac{c_B^b}{c_A^a} \tag{A.10}$$

where the c's denote concentrations.

The partition function is the sum of a series of terms of the type $e^{-\epsilon_i/kT}$, there being one term for each permitted energy level. Since $e^{-\epsilon_i/kT}$ is proportional to the number of particles in the i level it follows that Q is proportional to the total number of particles per unit volume, i.e. to the concentration. Thus the c's in eqn. (A.10) may be replaced by the corresponding Q's

$$K = \frac{(Q'_B)^b}{(Q'_A)^a} \tag{A.11}$$

The Q's in eqn. (A.11) have been given a prime because, in ensuring that the proportionality factor between Q and c is the same for reactants and products, the energy levels for A and B must be measured from the same zero, whereas the normal partition function of a species is referred to its own zero point level. It is desirable to write (A.11) in terms of normal partition functions. Suppose the zero point level of A, $\varepsilon_{0(A)}$, is $\Delta\varepsilon_{(A)}$ above a common zero and that of B, $\varepsilon_{0(B)}$, is $\Delta\varepsilon_{(B)}$ above the same zero. Then the energy levels in A referred to the new zero are $(\varepsilon_{0(A)} + \Delta\varepsilon_A)$, $(\varepsilon_{1(A)} + \Delta\varepsilon_A)$ and so on. The partition function with respect to this zero, Q_A' is

$$Q_A' = \Sigma_i e^{-(\varepsilon_{i(A)} + \Delta\varepsilon_A)/kT} \qquad (A.12)$$

$$= e^{-\Delta\varepsilon_A/kT} \Sigma_i \cdot e^{-\varepsilon_{i(A)}/kT} \qquad (A.13)$$

$\Sigma_i e^{-\varepsilon_{i(A)}/kT}$ is Q_A, the partition function referred to own zero point energy level as zero.
Therefore

$$Q_A' = e^{-\Delta\varepsilon_A/kT} \cdot Q_A \qquad (A.14)$$

Similarly for B

$$Q_B' = e^{-\Delta\varepsilon_B/kT} \cdot Q_B \qquad (A.15)$$

Hence

$$K = \frac{(Q_B)^b}{(Q_A)^a} \cdot \frac{e^{-(b \cdot \Delta\varepsilon_B)/kT}}{e^{-a \cdot \Delta\varepsilon_A/kT}}$$

$$= \frac{(Q_B)^b}{(Q_A)^a} e^{-(b\Delta\varepsilon_B - a\Delta\varepsilon_A)/kT} \qquad (A.16)$$

If the reaction took place at 0°K all the a particles of A would have energy $\varepsilon_{0(A)}$ and all b particles of $B \varepsilon_{0(B)}$. Thus $(b\Delta\varepsilon_B - a\Delta\varepsilon_A)$ is the change in internal energy accompanying the reaction if it took place at 0°K. Denote this change by $\Delta U°$. Then

$$K = \frac{(Q_B)^b}{(Q_A)^a} e^{-\Delta U°/kT} \qquad (A.17)$$

Factorization of Partition Functions

The complete partition function for a system contains terms for all possible combinations of the different types of energy possessed by the system. If the energy levels in each type are completely independent of the others then the complete partition function may be factorized into partial partition functions, one for each energy type. For example, suppose a system has rotational energy and energy of translation of the particles. Let the rotational energy levels be $\varepsilon_{0(R)}, \varepsilon_{1(R)}, \ldots, \varepsilon_{i(R)}$ and translation energy be $\varepsilon_{0(TR)}, \varepsilon_{1(TR)}, \ldots, \varepsilon_{i(TR)}$.

The rotational partition function is

$$Q_R = \Sigma_i e^{-\varepsilon_{i(R)}/kT} \tag{A.18}$$

and the translational partition function is

$$Q_{TR} = \Sigma_i e^{-\varepsilon_{i(TR)}/kT} \tag{A.19}$$

Provided the two sets of energy levels are independent the total energy of an atom is the sum of its rotational and translational energies. The first few combined levels are

$$\varepsilon_0 = \varepsilon_{0(R)} + \varepsilon_{0(TR)}; \quad \varepsilon_1 = \varepsilon_{0(R)} + \varepsilon_{1(TR)};$$
$$\varepsilon_2 = \varepsilon_{0(R)} + \varepsilon_{2(TR)}; \quad \varepsilon_i = \varepsilon_{0(R)} + \varepsilon_{i(TR)}$$
$$\varepsilon_0^1 = \varepsilon_{1(R)} + \varepsilon_{0(TR)}; \quad \varepsilon_i^1 = \varepsilon_{1(R)} + \varepsilon_{i(TR)}$$
$$\varepsilon_0^{11} = \varepsilon_{2(R)} + \varepsilon_{0(TR)}; \quad \varepsilon_i^{11} = \varepsilon_{2(R)} + \varepsilon_{i(TR)}$$

and there are levels for all possible combinations of the ε_R's and ε_{TR}'s. The complete partition function is

$$Q = \exp\left[\frac{-(\varepsilon_{0(R)} + \varepsilon_{0(TR)})}{kT}\right] + \exp\left[-\frac{(\varepsilon_{0(R)} + \varepsilon_{1(TR)})}{kT}\right] + \ldots$$
$$+ \exp\left[\frac{-(\varepsilon_{0(R)} + \varepsilon_{i(TR)})}{kT}\right] + \ldots$$
$$+ \exp\left[\frac{-(\varepsilon_{i(R)} + \varepsilon_{0(TR)})}{kT}\right] + $$
$$\ldots + \exp\left[\frac{-(\varepsilon_{1(R)} + \varepsilon_{i(TR)})}{kT}\right] + \ldots$$
$$= e^{-\varepsilon_0(R)} \cdot \Sigma_i e^{-\varepsilon_{i(TR)}/kT} + e^{-\varepsilon_1(R)} \Sigma_i e^{-\varepsilon_{i(TR)}/kT} + \ldots$$
$$\ldots + e^{-\varepsilon_{i(R)}} \Sigma_i e^{-\varepsilon_{i(TR)}/kT}$$
$$= \Sigma_i e^{-\varepsilon_{i(TR)}/kT} (e^{-\varepsilon_{0(R)}/kT} + \ldots + e^{-\varepsilon_{i(R)}/kT} + \ldots)$$

i.e.

$$Q = Q_{TR} \cdot Q_R \tag{A.19a}$$

Generalizing the result, the complete partition function is the product of all the partial partition functions. Further, the translational partition function, $Q_{(TR)}$, may be factorized into one for translation in each of three orthogonal axes x, y, z. That is

$$Q_{TR} = Q_{TR(x)} \cdot Q_{TR(y)} \cdot Q_{(z)} \quad \text{(A.20)}$$

and similarly for the other partition functions.

Translational Partition Function

If a particle of mass m moves freely in one direction (the x-direction) within a length l. Schrödinger's equation gives that

$$2p \cdot l = ih \quad \text{(A.21)}$$

where p is the momentum of the particle and i is the quantum number of the state.

Since the energy is pure kinetic energy

$$\varepsilon_{TR} = \tfrac{1}{2}m\dot{x}^2 = \frac{p^2}{2m} = i^2 \frac{h^2}{8ml^2}$$

$$Q_{TR(x)} = \sum_i e^{-i^2 h^2 / 8ml^2 kT}$$

The energy levels are very close so that they may be regarded as quasi-continuous allowing the summation to be replaced by an integral

$$Q_{TR(x)} = \int_0^\infty e^{-i^2 h^2 / 8ml^2 kT} \cdot di$$

This is a standard integral and it gives

$$Q_{TR(x)} = \frac{(2\pi mkT)^{1/2}}{h} l \quad \text{(A.22)}$$

Similar expressions hold for y and z directions. Thus the complete partition function for movement within a cubical box of side l is

$$Q_{TR} = \frac{(2\pi mkT)^{3/2}}{h^3} \cdot l^3 \quad \text{(A.23)}$$

Vibrational Partition Function

For a one-dimensional simple harmonic oscillator, quantum mechanics gives

$$\varepsilon = (i + \tfrac{1}{2})h\nu \qquad (A.24)$$

where ν is the frequency.

$$Q_{VIB} = \sum_i e^{-(i+\tfrac{1}{2})h\nu/kT}$$
$$= e^{-\tfrac{1}{2}h\nu/kT} \sum_i e^{-ih\nu/kT}$$

The summation is a geometric progression and so

$$Q_{VIB} = e^{-\tfrac{1}{2}h\nu/kT}(1 - e^{-h\nu/kT})^{-1} \qquad (A.25)$$

At high temperature $kT \gg h\nu$
and

$$Q_{VIB} \simeq kT/h\nu \qquad (A.26)$$

Appendix B. Units and Symbols

Two systems of units for the activation energy are in common use.

(a) U_A^0 is expressed in electron volts (eV) in which case the appropriate value of **k** in $e^{-U_A^0/kT}$ is 8.62×10^{-5} eV/°K.

(b) U_A^0 is given in calories per mole, the value of **k** being 1·98 cal/mole.°K. **k** is then the gas constant, usually denoted by **R**.

The conversion factor is $1 \, eV = 23{,}000$ cal/mole.

The following is a list of the important symbols used.

a_A	activity of the component A in a solution of A in B
a	lattice parameter of a cubic crystal
A	frequency factor in Arrhenius equation
A_s	temperature at which a martensite reaction reverses on heating
b	jump distance of a diffusing atom
c	concentration
d	interlamellar spacing of a eutectoid
D	diffusion coefficient
e	base of natural logarithms
E_A	empirical activation energy
f	correlation factor
F	Helmholtz free energy
G	Gibbs free energy
G_A	free energy of activation
$\bar{G}_A$	partial molar free energy of A in a solution
h	Plank's constants
H	enthalpy
i	general number
I	nucleation rate/unit volume
J	flux
k	Boltzmann constant

k	rate constant
K	equilibrium constant
Ln	natural logarithm
Log	logarithm to base 10
M	mobility of an atom
M_S	temperature of the start of a martensitic reaction on cooling
n	exponent of t in a Johnson–Mehl rate equation
N_A	atom fraction of A in an alloy A–B
N	number of atoms in a system
$\mathbf{N}$	Avogadros number
p	probability
Q_A	partition function of A
r	general radius
R	radius of growing precipitate
S	entropy
S_A	entropy of activation
t	time
T	temperature °K
U	internal energy
U_A	activation energy
W	free energy of formation of a critical nucleus
x	distance
y	fraction transformed
z	co-ordination number
α, β	phases
γ_A	activity coefficient of A in a solution
γ	surface energy
δ	disregistry
ε	energy level
ν	frequency of vibration
μ	shear modulus
λ	atomic volume
σ	strain energy/unit volume
Ψ_0	number of atoms captured at the interface of a growing crystal per unit area per unit time

Index

Activated state 6, 9
Activation energy 9, 11
  derivation 29
  empirical 30, 52, 53, 186
  for diffusion 70
  for interface reaction 162
  for nucleation 101
  for ring-diffusion 78
  for self-diffusion 78
  for solute-diffusion 80
  of diffusion controlled growth 186
  zero point 11
Activity 85
  coefficient 85
Arrhenius equation 22, 52
  deviations 33, 70
Austenite, decomposition of 34, 46, 52, 96
AUSTIN, J. B. 50, 52
  and Ricketts equation 50, 52
Autocatalysis 43, 208
AVRAMI, M. 145, 194

Bainite reaction 97
BECKER, R. 91, 101, 118, 122, 123, 130, 139
BILBY, B. A. 174, 176, 200
Boltzmann equation 12, 14
  constant 12
  distribution law 18
  factor 18
BORELIUS, G. 124, 127, 130
  theory of nucleation 124
BROOKS, H. 78, 140
BULLOUGH, R. 176
BURKE, J. 189

CAHN, J. W. 131, 143, 177, 183, 195
  and Hilliard model of nucleation 131
Catalysts 5, 7
  for nucleation 107, 112
C-curve *see* Time–temperature-transformation diagrams
Cellular precipitates 177
CHALMERS, B. 159
CHRISTIAN, J. W. 205
CLEMM, P. J. 143
Coarsening of precipitates 169
COHEN, M. 149, 202, 212
COMPAAN, K. 66, 78
Continuous precipitation 94, 99, 117, 146
Correlation factor for diffusion 66
  impurity diffusion 80
  vacancy diffusion 78
COTTRELL, A. H. 174, 176, 192
  Bilby equation 175
Critical nucleus 103, 123, 125, 130, 133, 136, 144
  two-dimensional 155
  work (free energy) of formation 103

DARKEN, L. S. 83, 84, 90
  equations 83
Degenerate states 18
DEHLINGER, U. 150
Diffusion 61
  chemical 65, 83
  enhancement by excess vacancies 82
  experimental studies of 67

INDEX

Diffusion (*contd.*)
  in cubic metals 71, 75
  interstitial solutes in b.c.c. metals 71
  mechanisms 75
  self 64
Diffusional transformations 93
  diffusion controlled growth of a fixed number of particles 185
  diffusion controlled growth of a variable number of crystals 188
  interface controlled growth 188
  kinetics 184
Diffusion coefficient 61
  determination of 67, 88
  for chemical diffusion 65, 84
  for tracer diffusion of components in homogeneous alloy 65, 86
  for (tracer) self-diffusion 64
  intrinsic chemical 86, 88
Diffusionless transformations *see* Martensitic transformations
Discontinuous precipitation 94, 99, 117, 145
Disregistry 137, 139
DÖRING, W. 101
Driving force for reaction 4
  diffusion 85
Droplet experiments 107, 108

Einstein model 19
Embryo 93
  free energy of formation 101, 133, 135
Empirical activation energy 30, 52
  determination of 53
Energy levels of oscillator 19
Entropy 1
  of activation 9, 12
  of activation, derivation of 30, 53
  of activation, empirical 31, 52
  partition function relationship 214
Equilibrium, conditions for 1
  constant 215
  metastable 1
  stable 2
  thermodynamics 5

Eutectoid reactions 92, 94, 177, 193

Fick's laws of diffusion 62, 86, 88
  Matano–Boltzmann solution 68
FISHER, J. C. 143, 211
Free energy
  change accompanying a martensitic reaction 203
  composition diagrams 118
  for iron–nickel system 202
  Gibbs 1
  Helmholtz 1, 214
  of formation of a critical nucleus 103
  of formation of an embryo 101
  of formation of an embryo on substrate 108
  of solid and liquid phases 105
Frequency factor 22
  determination 53
  empirical 52
  for diffusion 79
FUMI, G. 78

GREENWOOD, G. W. 171
Growth 45, 93
  controlled by diffusion 162, 185, 188, 191
  controlled by the interface process 161, 188
  dependent upon diffusion and interface process 167
  measurements of rate of 152
  of cellular precipitates 177
  of cylindrical precipitates 167
  of disc precipitates 167
  of eutectoids 177
  of martensitic plates 206
  of pearlite 178
  of precipitates from solution 32
  of single phase in single component system 154
  of single phase in two component system 160
  of spherical precipitate particles 162

Growth (contd.)
  stress assisted 172, 191
  theory of diffusional 152
Guinier–Preston zones 2, 34

Habit plane 137, 199
HAM, F. S. 165, 167, 176, 185, 192
HARDY, H. K. 115
HARPER, S. 175, 176
HAVEN, M. 66, 78
HEAL, T. J. 115
HILLERT, M. 48, 178
HILLIARD, J. E. 131
HINSHELWOOD, C. N. 20
HOBSTETTER, J. N. 130
  Scheil model for nucleation 130
HUNTINGTON, H. B. 78, 79
Hysteresis 103, 207

Impingement factor 46, 50, 166, 190, 193
Impurity diffusion 79
Incubation period 117
Interface 136
  between two regular solutions 122
  energy of austenite–martensite 212
  energy of coherent 137, 141
  energy of incoherent 133, 140
  energy of semi-coherent 140, 141
  energy of solid–liquid 107, 108
  incoherent 133
  semi-coherent 140
Interstitial
  activation energy for formation of 78
  activation energy for movement of 78
  mechanism of diffusion 75
  solution diffusion in b.c.c. metals 71
Isokinetic reactions 48, 55

JACKSON, K. A. 154, 159

JOHNSON, R. P. 81
  effect 81
JOHNSON, W. A. 46, 194
  and Mehl equation 46, 47, 51, 190, 193, 195

KAUFFMAN, L. 202
Kirkendall effect 70, 76, 77, 79, 84, 87

Lattice invariant deformation 199
LECLAIRE, A. D. 66, 80, 82
LOMER, W. M. 83

Martensite 34, 96, 196
  embryo 210, 212
  finish temperature (M_f) 206
  nucleation 209, 212
  reverse temperature (A_s) 207
  start temperature (M_s) 96, 97, 200, 212
  strain-nucleus model 212
  thermal stabilization 209
Martensitic transformations 95, 196
  antocatalytic effects 208
  athermal 206, 211
  crystallography 199
  effect of applied stress 201
  in gold–cadmium alloys 207
  in iron–nickel alloys 207, 212
  in iron–nickel–manganese alloys 207, 211
  in uranium–chromium alloys 208
  isothermal 207
  kinetics 206
  thermodynamics 199
Massive reaction 96
Matano interface 68, 88
Maxwell–Boltzmann distribution 18, 25
MEHL, R. F. 46, 177, 194
MISHIMA, T. 166
Mobility 85

NABARRO, F. R. N. 134, 136, 139

Negative-diffusion 90
NEWMAN, R. L. 176
Nucleation 45, 93, 98
 at dislocation 143
 at foreign surfaces 142
 at grain boundaries 142
 Borelius theory of 124
 Cahn–Hilliard model for 131
 catalysis 107, 112
 classical theory 100, 112
 classical theory in two-component systems 118
 coherent 136
 comparison of theory and experience 148
 experimental measurements of rate of 98
 from liquid 105, 107, 108
 heterogeneous 100, 142
 Hobsetter–Scheil model 130
 homogeneous 100
 incoherent 133
 influence of strain 132
 of graphite 149
 of martensite 209
 of pearlite 117, 177
 on impurities 108
 temperature dependence of maximum rate 114
 temperature dependence of rate 112
 two-dimensional 155
 variation with time 116, 144
 within spinodals 126, 132

Order of reaction 40
Orientation relationship 137, 199

Partition function 19, 214
 entropy relationship 214
 equilibrium constant relationship 215
 factorization 216
 for single harmonic oscillator 28, 219
 for translation 26, 218

Helmholtz free energy relationship 214
 internal energy relationship 215
Polymorphic transformations 92, 94, 154 193
Point defects, annealing out of 43, 58
 concentration of 77
POUND, G. M. 79
Precipitation from supersaturated aluminium alloys 2, 34, 82
 continuous 94
 discontinuous 94
 in gold–nickel alloys 149
 in gold–platinum alloys 123
 kinetics, growth only 189
 kinetics, nucleation and growth 191
 of cementite from ferrite 186
 solid solutions 92, 94

Quasi-equilibrium theory 22, 23
Quenching
 basis of 23
 of steel 96

Rate constant 26, 38
 dimensions of 40
 for first order reaction 41
 for Johnson–Mehl equation 46, 48, 50
 for second order reaction 42
 method of determining empirical activation energy and frequency factor 53
Rate equation 26, 38
 Austin–Rickett 50, 52
 for autocatalytic reaction 43, 44
 for first order reaction 40, 42
 for heterogeneous reaction 45
 for second order reaction 41, 42
Rate of reaction
 methods of determination 37
 reaction involving several processes 31
 singly activated reaction 21, 24

Reactions, endothermic 11
  bainite 97
  diffusional 93
  eutectoid 92, 94
  exothermic 11
  heterogeneous 3
  homogeneous 3
  martensitic 95
  massive 96
Relaxation time 41, 42
Retrogression *see* Reversion
Reversion 117
RICKETTS, R. L. 50, 52

SCHEIL, E. 130
SEITZ, F. 78
Shape change accompanying martensitic reaction 197
Shear transformation *see* Martensitic transformations
Sigmoidal reaction curves 45
Solidification 92, 154
  heterogeneous nucleation during 108
  isotherms 110, 111
  undercooling for 106
Solute-vacancy binding 81
Specific reaction rate *see* Rate constant
Spinodals 120
Stabilization (of martensite) 209
Strain energy of
  coherent embryo 139
  incoherent embryo 134
Stress-assisted precipitation 172

Thermal activation 7
Thomson-Freundlich equation 171

Time exponent (in rate equation) 192
Time-temperature-transformation diagrams 58, 115, 124, 128, 148, 208, 211
Transition phases 121, 146, 149
Transition state 6, 9
  theory 24
TURNBULL, D. 107, 109, 117

Undetermined multipliers, method of 16
Up-hill diffusion *see* Negative diffusion

Vacancy
  activation energy for formation 77, 78
  activation energy for movement 77, 78
  effect of non-equilibrium concentration 82
  mechanism of diffusion 75
  solute binding energy 81
Velocity constant *see* Rate constant
VOLMER, R. M. 101

WEBER, A. 101
WERT, C. 73, 164, 166, 167, 185, 186
Widmanstatten structures 141, 148

ZELDOVICH, I. 117
ZENER, C. 28, 30, 46, 73, 79, 164, 165, 167, 179, 185
Zero point energy 11